D1807594

Sustainable and Democratic Education

In a world struggling with environmental and social problems resistant to current solutions, education needs to explore ways to 'enlarge the space of the possible' rather than only 'replicate the existing possible'. To respond to this challenge, this book troubles dominant Western philosophical conceptions which continue to have wide-ranging influence in education worldwide and which limit more sustainable ways to be in the world together. It argues for the importance of opening spaces in and through which unique subjects can emerge, bringing potential for new ways of being and as yet unimagined futures.

The book makes a valuable contribution to international growing interest in Arendtian thinking, complexity and emergence, feminist thinking, the emerging field of anticipation studies, the posthuman and engagement with Indigenous scholarship and practices in ways which attempt to be non-appropriating. Sustainability continues to be a vital theme in education, and the book responds to a desire to encourage education which invites more sustainable processes and ways of being in addition to education which limits itself to teaching about, or for, sustainability.

Sustainable and Democratic Education will be of great interest to academics and practitioners working with sustainability, Indigenous scholarship, complexity theory and the posthuman and what these ideas can mean in and for education.

Sarah Chave is Associate Honorary Research Fellow at the University of Exeter, UK. She has researched and taught adult and higher education for over 30 years. Her areas of interest include teacher education, business and economics, language, embodied and arts-based research methodologies, equality, diversity and social justice and sustainability. Since completing her PhD in 2017, she has been involved in an international project researching decolonising teacher education and arts-based research looking at ways to develop more caring relationships with materials.

Routledge Research in Anticipation and Futures

Series editors: Johan Siebers and Keri Facer

The Promise of Nostalgia
Reminiscence, Longing and Hope in Contemporary American Culture
Nicola Sayers

Indigenous Futures and Learnings Taking Place
Edited by Ligia (Licho) López López and Gioconda Coello

Sustainable and Democratic Education
Opening Spaces for Complexity, Subjectivity and the Future
Sarah Chave

www.routledge.com/Routledge-Research-in-Anticipation-and-Futures/book-series/RRAF

Sustainable and Democratic Education

Opening Spaces for Complexity, Subjectivity and the Future

Sarah Chave

LONDON AND NEW YORK

First published 2021
by Routledge
2 Park Square, Milton Park, Abingdon, Oxon OX14 4RN

and by Routledge
52 Vanderbilt Avenue, New York, NY 10017

Routledge is an imprint of the Taylor & Francis Group, an informa business

British Library Cataloguing-in-Publication Data
A catalogue record for this book is available from the British Library

Library of Congress Cataloging-in-Publication Data
Names: Chave, Sarah, author.
Title: Sustainable and democratic education : opening spaces for complexity, subjectivity and the future / Sarah Chave.
Description: Abingdon, Oxon ; New York, NY : Routledge, 2021. | Series: Routledge research in anticipation and future studies | Includes bibliographical references and index.
Identifiers: LCCN 2020034921 (print) | LCCN 2020034922 (ebook) | ISBN 9780367151027 (hbk) | ISBN 9780429055027 (ebk)
Subjects: LCSH: Environmental education. | Sustainable development—Study and teaching. | Democracy—Study and teaching.
Classification: LCC GE70 .C4823 2021 (print) | LCC GE70 (ebook) | DDC 304.2071—dc23
LC record available at https://lccn.loc.gov/2020034921
LC ebook record available at https://lccn.loc.gov/2020034922

ISBN: 978-0-367-15102-7 (hbk)
ISBN: 978-0-429-05502-7 (ebk)

Typeset in Bembo
by Apex CoVantage, LLC

To Beatrice, my Mother

Contents

Figures

Tables

Acknowledgements

I would like to thank:

My teachers and friends, Deborah Osberg, Fran Martin, Alun Morgan and Alison Harper who have supported me generously in the writing of my PhD thesis and this book;

My husband Peter and our son James, for their unfailing support and patience;

The reviewers of my manuscript and the staff of Routledge for their help and feedback;

My mother Beatrice and grandfather Eddy, for their inspiration.

1 Enlarging the space of the possible

This book is about education which seeks to *enlarge the space of the possible* rather than replicate the existing possible (Davis *et al*. 2004). Such a reorientation is highly challenging at a time when education in many settings and across different age groups is focused on standardised knowledge and testing and the 'reining in' of imagination and creativity. However, this is a critically important move today, in the era of the Anthropocene, when many humans and their existing practices are changing the planet in permanent and damaging ways.[1] Inspired by Hannah Arendt, I argue in this book that opening opportunities for enlarging the space of the possible in education requires a move away from conceptions of education in which adults are the only ones 'who know' and towards a positioning in which education can be a place where:

> we decide whether we love our children [*and young people* – my addition] enough not to expel them from our world and leave them to their own devices, nor to strike from their hands their chances of undertaking something new, something unforeseen by us.
>
> (Arendt 2006 [1961]: 193)

In this book I argue that encounters are an important aspect of enlarging 'the space of the possible' (Davis *et al*. 2004: 4): encounters which encourage emergence of the 'radically new'[2] through opportunities for 'a fusion of mutual inspiration and experimentation' (Osberg 2019: 1472), and playful engagement with 'the boundless, incalculable possibilities of life which are not yet imagined or imaginable' (1472).[3] Such encounters open:

> an orientation to the future that admits of the possibility of future transformation that exceeds and resists colonisation by the constraints of the present.[4]
>
> (Facer 2016: 58)

It is important to recognise that the use of the word 'new' can appear as a denial of Indigenous ways of being in the world which have existed for millennia but which have been ignored or denigrated (see Kuhn1962 cited in St Pierre 2016, Todd 2016). The possibility of the 'radically new' explored in this book acknowledges and values these ways, desires to engage with them, and understands the radically new as what can emerge when different ontologies (ways of being) and epistemologies (ways of knowing) are valued and brought into conversation. This is not easy, due for example, to power imbalances and lack of awareness of one's engrained ways of framing the world. However, this book aims to contribute both to becoming aware of one's entrenched worldviews and to finding new ways to be and be together.

The 'education' particularly being called into question in this book and centred as problematic is one founded on entrenched Western (Eurocentric) philosophical traditions which foreground static conceptions of the world, separation (autonomy), and rationality. I recognise that Western philosophical ideas are not monolithic – there is variation within them which I discuss in Chapters 2 and 3. However, static conceptions of the world, separation (autonomy) and rationality are dominant in much Western thinking, and it is these conceptions that are challenged. Moreover, such dominant Western philosophical thinking is not geographically limited. It has spread throughout the world through processes such as colonialism and continues to have considerable influence worldwide on education. These ideas operate in ways unseen by many (but not all). They act as barriers to the emergence of other ways to be in and know the world and to possibilities for more sustainable and democratic education and new futures.

This book is undoubtedly theoretical, drawing on a range of thinkers including Arendt, Braidotti, Biesta, Little Bear, Andreotti, Todd, Donald, Gilligan, Topolski, Keller, Held, Noddings and Haraway. I draw on a wide variety of theoretical and philosophical debates to move forward ways of being, knowing and acting rather than offer in-depth introductions to each of the many thinkers and debates. The book is centred on education but also contributes to ongoing conversations that intersect and expand a range of fields of study including ecology, future studies, new materialism, the posthuman and politics. I acknowledge my own positioning in this process – the privileged and situated place I am speaking from – and explore this in more detail later in this chapter.

I would not wish, however, for the book to be understood as a development of theoretical ideas which are then to be 'tried out' in practice: a positioning which Doll (2008) describes as a habit of mind in which practice is seen as 'a handmaiden' to theory. This 'practice as handmaiden to theory' approach arises from scientific Western Modernist understandings which began to emerge in the Western world in the late Medieval and early Renaissance period in Europe and continued to develop in the centuries which followed (Doll 2008, Trueit 2005). These foreground reasoning about the world taking place before practice. This marked a transition from a previous discourse of '*patterning*' – a discursive, ongoing exchange between theory and practice in which both are held in esteem and inform each other. Drawing on a 'patterning' approach in education and in education research allows for ideas to develop in and through a discursive, recursive exchange between practice and theory: a *poietic* (creative) potential. Theoretical ideas shared in this book are not intended to sit outside of or precede practice. Instead the aim is that ideas can emerge in and through discursive exchange between theories shared, my experiences and teachers' and students' own experiences.

I include practical examples and case studies often drawn from my own experiences and encounters at the 'margins' of mainstream UK education where my own teaching life has been situated: places designed to support students who have not found conducive mainstream education dominated by particular Western philosophical ideas of how to understand and be in the world. The choice of examples reflects my commitment to recursive processes between my practice and the development of theoretical ideas. It also reflects the commitment in this book to the importance of first-hand encounters as a starting point for the opening up of spaces in and through which the new can emerge. The practical examples I include can help to elucidate the theoretical concepts shared, provide starting points for discussions and activities and also spark recollection and reflection of readers' and their students' experiences in their own educational contexts.

I begin this introductory chapter by exploring how complexity and emergence are necessary for sustainability, including the emergence of new subjectivities – new ways to be a human subject speaking and acting in the world. I discuss why the issue of subjectivity is an important aspect of education. I then outline future chapters and consider the lenses I bring to the book. I draw the chapter to a close by highlighting a role for hope in the challenging but important task of opening spaces for sustainable and democratic education in the era of the Anthropocene.

Sustainability and emergence

The possibility of emergence of the radically new introduces a way to examine and resolve the paradox which is inherent in the concept of 'sustainability'. To 'sustain' something is 'to cause it to continue for an extended period of time' (Oxford Living Dictionaries: English: online). The paradox is that to 'sustain' something – for something to 'continue' – it must also be allowed to *emerge* into something different than it currently is – an acorn into an oak tree, a baby into a ballet dancer. Moreover, it is not always knowable in advance what that 'something' will be. Such emergence is a key feature of ecological systems-thinking which recognises that systems do not remain static and unchanging over time. Adaptation and change are essential. Sand dune ecosystems consisting of minerals and a myriad of flora and fauna provide a helpful illustration (Adams 1996). 'Conservation' in this example involves finding a balance between the novelty naturally inherent in sand dune ecosystems over time and change introduced from outside these systems by humans. This can be too rapid for the dune ecosystems to cope with, leading them to be overwhelmed and to die. Ecosystems thinking acts to 'resolve' the paradox inherent in sustainability and allows sustainability to be understood as:

> The capacity to endure. In ecology the word describes how biological systems remain diverse and productive over time. For humans it is the potential for long-term maintenance of well-being which in turn depends on the natural world and natural resources.
>
> (Habitat.org.tr[5] 2016)

Approaches to environmental and sustainability education in dominant Western (Eurocentric) thinking show varying awareness of the importance of allowing the radically new to emerge. Sterling's framework of education *for*, *about* and *as* sustainability (Sterling 2003, 2008, 2009 [2001], 2010)[6] provides a helpful way to explore this issue of emergence. Education *about* sustainable development is first order learning covering knowledge and skills relating to sustainability issues which, as Sterling (2009 [2001]: 60) identifies, 'can be assimilated quite easily within the existing education paradigm'. Learning *for* sustainable development involves more reflective activities, encouraging the development of critical thinking skills; problem-solving and trans-disciplinary thinking which will be needed to solve sustainability issues in the future. However, as Sterling (2009 [2001]: 61) points out:

> There is often an assumption that we know clearly what values, knowledge and skills 'are needed'.

Sterling (2009 [2001]: 61) identifies a third approach – education *as* sustainability/ sustainable education. This is education which recognises the importance of emergence. It is a:

> transformative, epistemic learning response by the educational paradigm, which is then increasingly able to facilitate a transformative learning experience. This position subsumes the first two responses but emphasises process and quality of learning. . . . There is a keen sense of emergence and ability to work with ambiguity and uncertainty. Space and time are valued to allow creativity, imagination and cooperative learning to flourish.

Sterling (2009 [2001]) goes on to say that the existing educational paradigm is not set up in ways which can encourage such transformative, emergent processes. He argues that a paradigm shift in education (and here, I propose, he is referring to education informed by Western [Eurocentric] philosophical traditions) is therefore needed to encourage *sustainable education* which allows in and values emergence of the radically new.

Achieving such a paradigm shift in Western [Eurocentric]education is problematic. Whilst much has been written about education exploring the skills, knowledge and attitudes identified as necessary for sustainability/the environment, philosophical questions relating to *underpinning philosophical assumptions in education* itself have received less attention.[7] Writing as early as 1994, Jickling notes, 'One of the problems in environmental education has been the failure of its practitioners to reconcile definitions of environmental education with *a priori* conceptions of education' (1). This failure continues and contributes to the lack of the paradigm shift needed if education exploring the environment and sustainability is to make an impact in the world. One such neglected *a-priori* philosophical conception in education is *subjectivity,* which is assumed to be a particular thing, closing down explorations of other ways for it to be. Exploring subjectivity is a key focus of this book and it is to understanding subjectivity in more detail that I now turn.

A focus on subjectivity and opening possibilities for sustainable and democratic education

Education, including education exploring the issue of sustainability, is a multifaceted endeavour, engaging in many, *concurrent* aims and activities. Biesta (2010) highlights that two well-recognised aims of education are qualification (learning skills and knowledge) and socialisation (the insertion of newcomers into existing ways of knowing and being in the world). These activities occur concurrently so one learns knowledge and skills and also how these are valued and practiced in society (see Biesta 2006 for further discussion). The current technocratic approaches dominant in Western-influenced education emphasise these aspects of qualification and socialisation, focusing on 'what works' in relation to acquiring skills and knowledge and issues such as 'how do such learning and knowledge contribute to the economy'? Such questions can be important, especially at the practical level. However, there is an additional and sometimes overlooked aspect of education – *exploring what it is to be/become a (human) subject who acts in the world*. Simply put, in grammatical terms, the subject is the person or 'thing' doing an action in a phrase or sentence and the object is the recipient of the action. For example, in the phrase 'the man cuts down the tree' the man is the subject doing an action (in this case 'cuts down') and the tree is the object receiving the action. In philosophy, subjectivity refers to qualities that constitute the subject. In the Western philosophical tradition, 'subjectivity' is constructed as subjects having a sense of consciousness of the self: capacity to act and reflect on actions (such reflection is termed 'higher-order thinking') and having agency – the capacity that one has to act independently and make choices in the world. Western philosophy also emphasises autonomy and rationality as qualities of the subject.

In exploring this issue of subjectivity in education Biesta has developed the term 'subjectification'. Subjectification is the process of being/becoming an independent (human) subject who acts and is acted upon in the world (Biesta 2010, 2013). Subjectification does not replace existing education, rather it emerges in moments which interrupt, which challenge, existing framings and assumptions of subjectivity. 'Subjectification' can thus be understood as the opposite of socialisation and stands in tension with it. Biesta (2006) emphasises how education which *only* addresses qualification and socialisation runs the risk of being 'uneducational' since it denies the possibility of freedom for newcomers to bring new ways of knowing and new ways of being a subject into the world.

Biesta (2011) points to how the concept of *Erziehung* (one of the words used in the German language to refer to education) can help us to explore why engaging with possibilities of new ways to be a human subject – and the freedom for this to be conceivable – are important aspects of education. The term *Erziehung* came into the German language from the Reformation onwards and is concerned with the influences which impact on the soul of the human being. Initially explored in religious terms the notion of *Erziehung* was widened to include the secular scope of this issue. There have been various interpretations of the meaning and scope of *Erziehung*. Biesta (2011) cites for example the work of Hopmann (2007), Groothoff (1973) and Oelkers (2001). However, despite some differences they all share a theoretical concern with the issue of education as:

> an orientation toward emancipation, that is toward ways of doing and being that do not simply accept the given order but have an orientation toward the change of the existing order so that different ways of doing and being become possible.
>
> (Biesta 2013: 64)

As introduced earlier, the existing 'given order' that this book challenges is the dominant Western (Eurocentric) framing of subjectivity, and education informed by this positioning. This framing decides in advance that a subject is a rational autonomous individual, closing down possibilities of other ways of being in and knowing the world. The decision of '*who*' one is, or will become, is already made and education's role is to help in the development of such rational autonomous subjects. However, educational arrangements which leave the question of human subjectivity radically open create possibilities for other as yet unforeseen ways of being a subject to emerge. *The freedom which makes such emergence possible can be understood as an expression of democracy*: education as a radically democratic process. Such democratic education does not only include those excluded before, it is not just an issue of inclusion, albeit that is also important. It allows for a reconfiguration of how the world, and ways to be in it, are understood. As Facer (2016: 60) emphasises:

> It is in the reframing of democratic education as a politics of disclosing and holding open new possibilities *for all future generations* [original italics]

rather than the realising of the pre-defined dreams of today's generation of adults, that a different form of educational project becomes available.

An orientation towards the future informed by Facer's (2016: 58) 'pedagogy of the present' – a resisting of colonisation of the future by ideas of the present – thus allows for 'the possibility for novelty in the future' (58), and for a 'rich possibility of *different* ways of being' (58). Such an approach goes beyond the epistemological problem that someone cannot know the future since it has not yet happened. Instead it is an:

> ontological assertion, that the future will constitute a different reality, that it will bring ways of being, of living, of knowing, that are different from today and from the past.
>
> (Facer 2016: 59)

In such an orientation 'the educational encounter is understood as constituting a distinctive temporality of its own' (58). Such temporality can be encouraged by:

> the intentional putting into play of the abundant materials that constitute and create futures, that keeps them intentionally open for exploration, and that expands and encourages the space to participate within that process.
>
> (58–59)

Education as a potentially radically democratic process in the ways explored here – *democracy understood as freedom to open spaces in and through which unforeseen ways of being and living in the world can emerge* – can also be understood as education as a sustainable process. Through challenging framings currently dominant but often unacknowledged in Western (Eurocentric) education such democratic education opens possibilities for the paradigm shift called for by Sterling if education is to become sustainable education (2009 [2001]).

Chapter outlines

Encouraging educational encounters which open spaces in and through which unforeseen ways to be a subject can emerge requires recognising and challenging current dominant Western (Eurocentric) framings of the world and of subjectivity. I therefore begin Chapter 2 by highlighting current dominant Western (Eurocentric) static conceptions of the world, made up of fundamentally separate parts. I explore the philosophical roots of these ideas and consider why they are problematic, in particular in relation to the possibility of the emergence of the radically new. I then introduce and explore ideas from Heraclitus, Bergson, complexity theory and Indigenous thinking, which provide ways to understand the world as a place of flux and emergence, including why these approaches are important and timely for education engaging with issues of sustainability and democracy. I acknowledge the risk of 'appropriation' and grafting of Indigenous knowledges onto dominant Western framings without understanding and engaging with their full meaning. I highlight the harms that dominant ontologies have caused, and continue to cause to others in the world (both human and other-than human) and also the continuing influence of one's own framings when engaging with the framings of others (see Bonnett 2002, Stables and Scott 2001, Andreotti 2012, 2016, Todd 2016). I also consider here that what some describe as 'new' are often ancient ontologies and epistemologies which have been unrecognised or denigrated.

In Chapter 3 I foreground how dominant Eurocentric understandings frame subjectivity in terms of autonomy and rationality. Drawing on feminist theories of care and Indigenous knowledges I then consider the possibility of other ways to be a (human) subject in the world. I introduce these ideas, not as 'blueprints' or 'models' for subjectivity. Rather, they are included to show that other ways of being *are possible* and can be brought into dialogue with existing dominant framings. I recognise that incommensurabilities exist in these different ideas, and these need to be respectfully recognised alongside the possibilities opened up by seeing what emerges when these 'other(ed)' ideas and dominant Western ideas are brought into conversation.

In Chapter 4 I focus on the issue of Western notions of human subjectivity in more depth. I begin by considering the terms 'subject' and 'subjectivity', as well as issues that thinking about (theorising) subjectivity creates. I then explore the ideas of two very different contemporary thinkers: Biesta's (2006, 2010, 2013) 'pedagogy of interruption' and Braidotti's (2011a, 2011b, 2013) 'nomadic subject'. I consider how my thinking both draws on and departs from these ideas, as well as reflecting on engagement with the ideas explored in Chapters 2 and 3. I then introduce and discuss Arendt's thinking on the *vita activa*, including her conception of 'spaces of appearance', 'natality' and *'potentia'* and how these can contribute to opening spaces for the emergence of new subjectivities.

In Chapter 5 I explore ways which have the potential to encourage (but never guarantee) the possibility of opening of 'spaces of appearance' in which new subjectivities can emerge. The chapter draws further on the ideas of Arendt introduced in Chapter 4 and also ideas of a variety of European thinkers including Mouffe's (2000, 2005) agonistic pluralism, Rancière's (2010a) conceptions of *dissensus* and stultification and Masschelein and Simons' (2013) ideas of education as *skholé*. Reading these authors together could be considered a controversial move, especially since both Mouffe and Rancière have openly critiqued Arendt's work. However, I draw on Dikeç (2013: 78) to argue that such an endeavour is worthwhile since it brings together ways to think about the possibility of an education which is both 'ruptural and inaugurative'. I recognise that encouraging conditions in which emergence of the new is a possibility raises important ethical concerns and challenges and therefore indicate that ethics is the focus of Chapter 6.

In Chapter 6 I propose Arendt's framework of forgiveness and mutual promising as a way to address the ethical issues arising from the irreversibility, unexpectedness and unboundedness of the radically new. In making this argument I include Topolski's (2011) emphasis on what she calls the 'Judaic' in Arendt's thinking and the importance of acts of turning or '*shuv*' in developing an immanent approach to ethics. I include a range of case studies to explore and provide examples for use of these ideas in educational settings. Included here are discussions of Arendt's thinking on how extreme evil develops and the issue of possible limits to forgiveness. I also make connections to Indigenous thinking and to feminist ethics of care introduced in Chapters 2 and 3.

In Chapters 7 and 8 I develop a concept of intersubjective encounters with humans and other participants in the wider natural world of which humans are a part. I argue that such encounters have potential to open 'spaces of appearance' in which *who* one is can emerge intersubjectively, thus inserting new ways of knowing, being and acting in the world we share. In such intersubjective encounters, happening through expressive enactments and 'entanglements', one does not reveal an inner essence to others. Instead, in such encounters, 'spaces of appearance' (Arendt 1974 [1958]) can open up through which a unique subject can emerge, arising in what Topolski (2015: 176) calls the '*the space between I and we*'. Engaging in this space, neglected in dominant Western philosophy, has the potential to create a surplus. Something radically new can emerge, something that was not in the world before and which 'cannot be predicted from the ground from which it arose' (Osberg and Biesta 2008: 313): bringing potential to open up unforeseen, more sustainable ways to be, and be together, in the world and to resist colonisation of the future by ideas and attitudes of the present.

A personal introduction: the lenses I bring to the ideas explored in the book and my motivation for writing it

As Nagel (1989) reminds us, all ideas are a view from 'somewhere' despite attempts in much writing to position one's research and thinking as objective and disconnected from the researcher. In writing this book I acknowledge and celebrate that I am unavoidably part of the writing I produce. It therefore seems fair that I should share with you a little about myself and the lenses I bring.

My son recently asked me 'why did you get so interested in sustainability?' I found I was able to respond readily to this question. For me, education is intrinsically connected with thinking about the future. Younger students are in education before entering their future lives – and we have a responsibility both to think about and care for the future world they will be inheriting, as well as trying to prepare them for such futures. Yet this is no easy task: we do not know what the future will hold, so how to help to prepare for it? As my interest in sustainability grew so too did my dissatisfaction with many books on sustainability as they often seemed to me to be very directional – do this, but don't do that. As Sterling (2009 [2001]: 61) puts it, 'There is often an assumption that "we" know clearly what values, knowledge and skills "are needed" to do things better and also what those "better things" might be'. Since my 'teaching motto' has always been 'ask, don't tell', or at least to allow some time for this, it seemed to me that such an 'open' approach should also be allowed into education exploring the issue of sustainability. It felt important to provide opportunities for students to engage with what Poli (2011 cited in Facer 2016: 58) calls 'the dynamism and emergent properties of the thick present'.

Whilst researching sustainability, alongside teaching education studies, economics and business studies in a UK further (vocational) education college I encountered Vare and Scott's (2007: 1) model of ESD1 and ESD 2 (ESD is an abbreviation of 'education for sustainable development'). In this model ESD1 includes education *about* sustainability issues e.g. education on the topic of climate change, and education *for* sustainability i.e. education promoting ways to be more sustainable now and in the future. They proposed that education *for* and education *about* sustainability form one side of an interconnected whole. The other side of this whole is ESD2 – 'building capacity to think critically about what experts say and to test ideas, exploring the dilemmas and contradictions inherent in sustainable living'. Both aspects are valuable. Vare and Scott recognised the importance of ESD1 approaches but felt that ESD2 approaches were often lacking in education (and here they are referring to Western education) exploring sustainability and that this was an area which needed further research and ideas. I also felt this, and therefore this is where I locate my research and thinking.

As my research progressed, I became increasingly interested in the issue of subjectivity – the 'I' acting in the world. As already noted earlier in this chapter the issue of subjectivity is often either unexamined, an unspoken assumption in Western education. If considered at all, then it is taken-for-granted that subjectivity based on rationality and autonomy is a natural state of affairs (this is explored in detail in Chapter 3) and education's role is to produce, or in a more facilitative mode 'allow', the development of rational, autonomous individuals. For me, this raised the question of whether there were other ways to be a subject, ones foregrounding, for example, emotion, responsiveness to the other, interconnection or indeed other ways as yet unimagined. For me this seemed important for sustainability since it opens possibilities for new ways to be together in the world. Engaging with other subjectivities also seemed an important democratic move where democracy is understood as an expression of freedom – freedom to choose who to be and become.

Researching this book also highlighted for me the challenges of engaging with Indigenous knowledges when not speaking from within these particular people, places and traditions. An important issue is the dangers of appropriation – drawing on ideas when it suits Western societies after centuries of othering and denigrating such ideas. I identified with McIntosh's (1989: 10) observation that she 'was taught to see racism only in individual acts of meanness, not in invisible systems conferring dominance on my group', an invisible privilege often called 'white privilege' in academic literature. 'White privilege' can be understood as:

> an invisible package of unearned assets I can count on cashing in each day but about which I was meant to remain oblivious. White privilege is like an invisible weightless knapsack of special maps, passports, codebooks, visas, tools and blank cheques.
>
> (10)

I learnt about acknowledging the systemic 'white privilege' I have benefited from in my life; the necessity of recognising the systemic privileges afforded to some and not others and the historic *and* ongoing harm the conferring of privilege in this way causes. This is important if processes of engaging with Indigenous ideas are to go beyond further taking from 'othered' groups, individuals and the wider natural world: a continuation of a colonial mindset. Todd suggests Sundberg's (2014) toolkit (discussed in detail in Chapter 2) to approach these issues of appropriation and systemic 'white privilege'. Sundberg proposes a process which draws first on what Spivak (1993) calls 'homeworking' to identify one's existing ontological and power positionings and the implications of these. Second she introduces the Zapatista movement's concept of *preguntando caminamos* – 'asking as we walk' – which emphasises ways to 'learn from', 'learn with' and develop respect for a multiplicity of 'others' rather than being driven by a 'will to know', a will to categorise and to own.

However, there is another dimension to my experiences in relation to exploring Indigenous knowledge. My father was what was called 'Anglo-Burmese'. His mother was part of the Anglo-Burmese community, and his father was a British district judge/civil servant in Burma. The journalist Sue Arnold (1996), who has the same sort of family history, has written about the Anglo-Burmese community and the widespread practice in the British Empire in Burma of having a local 'Burmese family' if the British wife and family did not accompany the colonial office holder. The Burmese family was established in a colonial household with staff and was socially separated from the wider community. On the UK television programme *Who do you think you are?* (2014), the presenter and DJ Reggie Yates learnt about his own family history in Ghana where there was a similar system. When the 'tour of duty' was over, or if the British family joined him, the British father left his 'local' family. Commonly, no more contact was made despite causing issues around abandonment for those left behind: issues which people then took forward into their own adult relationships and the relationships they had with their own children. I felt a sense of solidarity with the words of Arnold and Yates, a sense of inhabiting an in-between-space between the coloniser and the colonised: a space rarely spoken about or even acknowledged in both family and national histories.

My personal experience of colonisation is a lens I bring to the writing of this book. I am writing from a place where colonisation is not a binary and, for me, connection to other ways of being a human subject in the world is both a possibility and part of my own historical and lived identity. De Sousa Santos (2007: 1) identifies what he calls 'abyssal' thinking which:

> operates through radical lines that divide social reality into two realms, the realm of 'this side of the line' and the realm of 'the other side of the line'. The division is such that 'the other side of the line' vanishes as reality, becomes non-existent, and is indeed produced as non-existent. What most fundamentally characterizes abyssal thinking is thus the impossibility of the copresence of the two sides of the line.

My lived experience has given me glimpses across this abyss, to see the possibility and value of the 'copresence of two sides of the line', and confidence to challenge often hidden assumptions in Western education.

Sustainable and democratic education: a role for hope

As I have explored in this opening chapter, engaging with new ways to be a subject who acts and takes responsibility in the world creates a bridge between education *as* radically democratic process and education *as* sustainable process. Such education goes beyond activism in existing framings of the world. Instead sustainable and democratic education can be understood as an *educational praxis*: an enabling of what Smith (2011: online) calls 'action by people who are free', who are able to both reflect and act as unique beings in particular encounters in ways which are open to the other and which are creative and emergent.

Whilst finding such new ways to be, and to be together, in this world we share is a challenge, it is a challenge which can be approached with reciprocity and with hope. However, this is not hope understood as some kind of passive 'wishful thinking'. Rather, to cite Bloch (1995 [1954]: 1):

> Hope, superior to fear, is neither passive like the latter, nor locked into nothingness. . . . The work of this emotion requires people who throw themselves actively into what is becoming, to which they themselves belong.

The educator and environmentalist David Orr's (2007: 1392) assertion that 'hope is a verb with its sleeves rolled up' resonates with Bloch's thinking. It is on such active conceptions of hope that this book draws.

The current crisis the world faces in relation to climate change, species loss and environmental degradation is not only an environmental crisis – it is a crisis of 'Reason' itself (Plumwood 2001, Rigby 2019). In an era dominated by Western (Eurocentric) notions of 'Reason' and conceptions of the autonomous individual existing in a static world, certain humans are destroying the very source of their own survival as well as the survival of others, both human and other-than-human. This is a situation striking in its 'unreasonableness'! The domination of such Western worldviews needs to be challenged. Other framings, and the hopeful opportunities that engaging with these can open up, need to be considered and these are therefore the focus of Chapters 2 and 3.

Notes

1 The 'Anthropocene' is a term used widely since its coining by Paul Crutzen and Eugene Stoermer in 2000 to denote the present time interval, in which many geologically significant conditions and processes are altered profoundly by human activities. These include: changes in erosion and sediment transport associated with a variety of anthropogenic processes, such as colonisation, agriculture, urbanisation and global warming, the chemical composition of the atmosphere, oceans and soils, with significant anthropogenic perturbations of the cycles of elements such as carbon, nitrogen, phosphorus and various metals. Environmental conditions generated by these perturbations include global warming, ocean acidification and spreading 'dead zones' in the biosphere both on land and in the sea, as a result of habitat loss, predation, species invasions and physical and chemical changes (International Commission on Stratigraphy [ICS] 2016). In September 2016, the ICS working party voted in favour of accepting the term subject to identification of a specific signal which can mark the change. It is now undertaking further research to establish whether such a signal can be identified (Carrington 2016).
2 In this book I use the term 'radically new' to indicate something 'which is uniquely new, something which has not been in the world before, and cannot be predicted from the ground from which it emerged' (Osberg and Biesta 2008: 313). This is discussed in detail in Chapter 2.
3 These ideas are part of Osberg's (2019) conception of 'symbiotic anticipation' which I return to in chapter 5.
4 Facer develops these ideas in her 'pedagogy of the present' (Facer 2016: 58).
5 Habitat.org.tr is a non-profit organisation in Turkey. Habitat is founded for research and publishes information about human settlements, environment, education and health. Their website is: http://habitat.org.tr/
6 See also Bonnett (2000, 2002), Foster (2001, 2011), Orr (2004), Vare and Scott (2007).
7 Educationists who have engaged with metaphysical issues include Bonnett (2000, 2002, 2004, 2012), Bonnett and Cuypers (2002), Foster (2001, 2011), Stables (2006), Stables and Gough (2006).

References

Adams, W.M. (1996) *Future matters: A vision for conservation*. Abingdon, Oxon: Routledge.
Andreotti, V. (2012) Editor's preface: HEADS UP. *Critical Literacy: Theories and Practices*, 6(1): 1–3. Available at: www.oregoncampuscompact.org/uploads/1/3/0/4/13042698/andreotti_-_preface_-critical_literacy_org_-_headsup__1_.pdf [Accessed 23.1.2020].
Andreotti, V. (2016) Multilayered selves: Colonialism, decolonization and counter-intuitive learning spaces. *Arts Everywhere, Musagetes*. Available at: http://artseverywhere.ca/2016/10/12/multi-layered-selves/ [Accessed 11.8.2018].
Arendt, H. (1974 [1958]) *The human condition*. Chicago, IL: University of Chicago Press.
Arendt, H. (2006 [1961]) Crisis in education. In *Between past and future: Eight exercises in political thought*. London: Penguin.
Arnold, S. (1996) *A Burmese legacy*. London: Coronet Books.

Biesta, G. (2006) *Beyond learning: Democratic education for a human future*. Boulder, CO: Paradigm Publishers.

Biesta, G. (2010) *Good education in an age of measurement: Ethics, politics, democracy*. London: Paradigm Publishers.

Biesta, G. (2011) Disciplines and theory in the academic study of education: A comparative analysis of the Anglo-American and Continental construction of the field. *Pedagogy, Culture & Society*, 19(2): 175–192. doi: 10.1080/14681366.2011

Biesta, G. (2013) *The beautiful risk of education*. London: Paradigm Publishers.

Bloch, E. (1995 [1954]) *The principle of hope, vol. I*. Cambridge, MA: MIT Press.

Bonnett, M. (2000) Environmental concerns and the metaphysics of education. *Journal of the Philosophy of Education*, 34(4): 591–602.

Bonnett, M. (2002) Education for sustainability as a frame of mind. *Environmental Education Research*, 8(1): 9–20. doi: 10.1080/13504620120109619

Bonnett, M. (2004) *Retrieving nature: Education for a post-humanist age*. London: Wiley.

Bonnett, M. (2012) Environmental concern, moral education and our place in nature. *Journal of Moral Education*, 43(1): 285–300.

Bonnett, M. and Cuypers, S. (2002) Autonomy and authenticity in education. In N. Blake, P. Smeyers, R.D. Smith and P. Standish (eds.) *The Blackwell guide to the philosophy of education*. Oxford: Wiley-Blackwell, pp. 326–340.

Braidotti, R. (2011a) *Nomadic theory: The portable Rosi Braidotti*. New York: Columbia University Press.

Braidotti, R. (2011b) *Nomadic subjects: Embodiment and sexual difference in contemporary feminist theory*, 2nd Edition. New York: Columbia University Press.

Braidotti, R. (2013) *The posthuman*. Boston, MA and Cambridge: Polity Press.

Carrington, D. (2016) The Anthropocene epoch: Scientists declare dawn of human-influenced age. *The Guardian*, 29th August. Available at: www.theguardian.com/environment/2016/aug/29/declare-anthropocene-epoch- experts-urge-geological-congress-human-impact-earth [Accessed 10.11.2016].

Davis, B., Phelps, R. and Wells, K. (2004) Complicity: An introduction and a welcome. *Complicity: An International Journal of Complexity and Education*, 1(1): 1–7.

De Sousa Santos, B. (2007) Beyond abyssal thinking: From global lines to ecologies of knowledge. *Revista Crítica de Ciencias Sociais,* 30(1).

Dikeç, M. (2013) Beginners and equals: Political subjectivity in Arendt and Rancière. *Transactions of the Institute of British Geographers*. Royal Geographical Society (with the Institute of British Geographers), 38(1): 78–90.

Doll, W.E. Jr. (2008) Complexity and the culture of curriculum. *Educational Philosophy and Theory*, 40(1): 190–212. doi: 10.1111/j.14695812.2007.00404.x

Facer, K. (2016) Using the future in education: Creating space for openness, hope and novelty. In H.E. Lees and N. Noddings (eds.) *The Palgrave international handbook of alternative education*. London: Palgrave Macmillan, pp. 53–78.

Foster, J. (2001) Education *as* sustainability. *Environment Education Research*, 7(2): 153–165. doi: 10.1080/13504620120043162

Foster, J. (2011) Sustainability and the learning virtues. *Journal of Curriculum Studies*, 43(3): 383–402. doi: 10.1080/00220272.2010.521260

Groothoff, H.-H. (1973) Theorie der Erziehung. In H.-H Groothoff (ed.) *Pädagogik Fischer Lexikon*. Frankfurt am Main: Fischer Taschenbuch Verlag, pp. 2–79.

Habitat.org.tr (2016) *Sustainability*. Available at: http://environment-ecology.com/what-is-sustainability/247-sustainability.html [Accessed 12.11.2016].

Hopmann, S. (2007) Restrained teaching: The common core of Didaktik. *European Educational Research Journal*, 62: 109–124.

International Commission on Stratigraphy (2016) Subcommission on Quaternary Stratigraphy: *Working group on the 'Anthropocene'*. Available at: http://quaternary.stratigraphy.org/workinggroups/anthropocene/ [Accessed 10.11.2016].

Jickling, B. (1994) Why I don't want my children to be educated for sustainable development: Sustainable belief. *Trumpeter*, 11(3). Available at: http://trumpeter.athabascau.ca/index.php/trumpet/article/view/325/497 [Accessed 1.1.2017].

Kuhn, T.S. (1970 [1962]) *The structure of scientific revolutions*, 2nd Edition. Chicago, IL: University of Chicago Press.

Masschelein, J. and Simons, M. (2013) *In defence of schools: A public issue*. Leuven: Education, Culture and Society Publishers. Available at: http://ppw.kuleuven.be/ecs/les/in-defence-of-the-school/masschelein-maarten-simons-in-defence-of-the.html [Accessed 3.2.2016].

McIntosh, P. (1989) White privilege: Unpacking the invisible knapsack. *Peace and Freedom Magazine: A Publication of the Women's International League for Peace and Freedom*, Philadelphia, PA, July/August, pp. 10–12.

Mouffe, C. (2000) *Deliberative democracy or agonistic pluralism* (Political science series). Vienna: Institute for Advanced Studies. Available at: www.ihs.ac.at/publications/pol/pw_72.pdf [Accessed 2.12.2016].

Mouffe, C. (2005) *The democratic paradox*. New York: Verso.

Nagel, T. (1989) *The view from nowhere*. Oxford: Oxford University Press.

Oelkers, J. (2001) *Einführung in die Theorie der Erziehung* (Introduction to educational theory). Weinheim and Basel: Beltz.

Orr, D. (2004) *Earth in mind: On education, environment, and the human prospect*, 2nd Edition. Washington, DC: Island Press.

Orr, D. (2007) Optimism and hope in a hotter time. *Conservation Biology*, 21(6): 1392–1395. doi: 10.1111/j.1523-1739.2007.00836.x

Osberg, D. (2019) Education and the future: Rethinking the role of anticipation and responsibility in multicultural and technological societies. In R. Poli (ed.) *Handbook of anticipation*. Cham, Switzerland: Springer International, pp. 1459–1478.

Osberg, D. and Biesta, G. (2008) The emergent curriculum: Navigating a complex course between unguided learning and planned enculturation. *Journal of Curriculum Studies*, 40(3): 313–328. doi: 10.1080/00220270701610746

Oxford Living Dictionaries: English (online) Sustain. *Oxford Living Dictionaries: English*. Available at: https://en.oxforddictionaries.com/definition/sustain [Accessed 11.1.2017].

Plumwood, V. (2001) *Environmental culture: The ecological crisis of reason*. London: Routledge.

Poli, R. (2011) Steps towards an explicit ontology of the future. *Futures*, 16(1), 67–78.

Rancière, J. (2010a) Who is the subject of the rights of man? In J. Rancière (ed.) *Dissensus: On politics and aesthetics* London: Continuum, pp. 70–83.

Rigby, K. (2019) "Piping in their honey dreams": Towards a creaturely ecopoetics. In F. Middelhoff, S. Schönbeck, R. Borgards. and C. Gersdorf (eds.) *Texts, animals, environment: Zoopoetics and ecopoetics*. Freiberg: Rombach Verlag, pp. 281–295.

Smith, M.K. (2011) What is praxis? In *The encyclopedia of informal education*. Available at: http://infed.org/mobi/what-is-praxis/ [Accessed 27.9.2019].

Spivak, G. (1993) *The post-colonial critic: Interviews, strategies, dialogues*. New York: Routledge.

Stables, A. (2006) *Living and learning as semiotic engagement: A new theory of education*. New York: Mellen Press.

Stables, A. and Gough, S. (2006) Toward a semiotic theory of choice and of learning. *Education Theory*, 56(3): 271–285. doi: 10.1111/j.1741-5446.2006.00226.x

Stables, A. and Scott, W. (2001) Post-humanist liberal pragmatism? Environmental education out of modernity. *Journal of Philosophy of Education*, 35(2): 269–279. doi: 10.1111/ 1467-9752.00225

Sterling, S. (2003) *Whole systems thinking as a basis for paradigm change in education: Explorations in the context of sustainability* (Unpublished PhD thesis). University of Bath, Bath. Available at: http://www.bath.ac.uk/cree/sterling/sterlingthesis.pdf [Accessed 24.9.2020].

Sterling, S. (2008) Sustainable education – Towards a deep learning response to unsustainability. *Policy & Practice: A Development Education Review*, 6, Spring. Available at: https:// www.developmenteducationreview.com/issue/issue-6/sustainable-education-towards-deep-learning-response-unsustainability [Accessed 7.1.2017].

Sterling, S. (2009 [2001]) *Sustainable education: Re-visioning learning and change*. Dartington: Green Books Ltd for the Schumacher Society.

Sterling, S. (2010) Transformative learning and sustainability: Sketching the conceptual ground. *Learning and Teaching in Higher Education*, 5: 17–33. Available at: http://www2. glos.ac.uk/offload/tli/lets/lathe/issue5/Lathe_5_S%20Sterling.pdf [Accessed 7.1.2017].

St Pierre, E. (2016) Curriculum for new material, new empirical enquiry. In N. Snaza, D. Sonu, S.E. Teruman and Z. Zaliwska (eds.) *Pedagogical matters: New materialisms and curriculum studies*. New York: Peter Lang.

Sundberg, J. (2014) Decolonizing posthumanist geographies. *Cultural geographies*, 21(1): 33–47.

Todd, Z. (2016) An Indigenous feminist's take on the ontological turn: 'Ontology' is just another word for colonialism. *Journal of Historical Sociology*, 29(1): 4–22.

Topolski, A. (2011) The ethics and politics of *Teshuvah*: Lessons from Emmanuel Levinas and Hannah Arendt. *The University of Toronto Journal of Jewish Thought*, 2. Available at: http:// tjjt.cjs.utoronto.ca/wp-content/uploads/2013/11/Anya-Topolski-The-Ethics-and-Politics-of-Teshuvah-Lessons-from-Emmanuel-Levinas-and-Hannah-Arendt-JJT-Vol.-2.pdf [Accessed 4.11.2015].

Topolski, A. (2015) *Arendt, Levinas and a politics of relationality* (Reframing the boundaries: Thinking the political). London: Rowman & Littlefield International.

Trueit, D.L. (2005) *Complexifying the poetic: Toward a poiesis of curriculum*. Available at: http:// etd.lsu.edu/docs/available/etd-11152005-184410/unrestricted/Trueit_dis.pdf [Accessed 28.6.2016].

Vare, P. and Scott, W. (2007) Learning for a change: Exploring the relationship between education and sustainable development. *Journal of Education for Sustainable Development*, 1(2): 91–198.

Who do you think you are? (2014) Reggie Yates. Series 11, episode 8. *BBC1*. Thursday 25th September 21.00.

2 Recognising and challenging static framings and ways of being in the world

The river
where you set
your foot just now
is gone – those waters
giving way to this,
now this.
(Fragment 44, Heraclitus 500 BCE – discussed on page 20)

Introduction

Looking out of the window, greeting a friend, lifting a glass, thinking about life – these are actions one engages in without necessarily stopping to consider, to feel, how one's understanding of the world frames those experiences. Indeed, one can be so accustomed to these framings – these 'worldviews' – that they appear to be just 'the ways that things are'. Worldviews are sometimes, especially in philosophical discourses, called 'metaphysics'. Leroy Little Bear (2016: online) explains metaphysics as:

> the foundational bases of our thinking processes. They are the criteria used on a moment-to-moment, day-to-day basis and they are so embodied in us that they go unquestioned, unstated and taken for granted but form the basis of our beliefs, what we do and how we relate.

It is, however, possible to take a 'step back' and to develop one's capacity *to see* one's worldview rather than to *see with* one's worldview (Sterling 2010). In this chapter I identify and explore how dominant Western (Eurocentric) worldviews (metaphysics) frame the world as fundamentally stable, made of separate, orderly, unchanging parts. I examine the origins of these often-hidden dominant ways of thinking about the world, and connections within it, and the *problems* that these ways of thinking can create in education, particularly in education which wishes to take the issues of sustainability and democracy seriously. Dominant Western philosophical framings are the focus for analysis and challenge since they have a widespread influence, often in unseen ways, in education throughout the world. These are due to processes such as colonialism and the ongoing structures that colonialism introduced which still continue today. I reflect on why challenging these understandings is important in this era of the Anthropocene as the world faces unprecedented threats such as mass species loss (IPBES 2019) and climate change (IPCC 2014, 2018). I then explore other ways of framing the world, with a particular focus on what such exploration can contribute to sustainable and democratic education.

Exploring dominant Western (Eurocentric) framings of the world

A helpful starting point for exploring and critiquing dominant Western (Eurocentric) framings of the world lies in the writings of Ancient Greeks, especially those of Parmenides and Heraclitus. Graeber (2001: 50) highlights how in Parmenides' thinking 'for objects to be comprehensible they must exist to some extent outside time and change' and that 'there is a level of reality, perhaps one that humans can never fully perceive, at which forms are fixed and perfect'. This Parmenidean view became dominant in Hellenic (Greek) thinking and, through the ideas of Pythagoras and Plato, still continues to be very influential on Western (Eurocentric) mathematical, scientific and philosophical traditions and understandings of the world today.

It is important to note here that Western philosophical conceptions are not monolithic. They have been challenged both from within and through encounters with other philosophical and religious traditions. For example, internal challenges to the Western European Enlightenment (a key movement in Western philosophy which is discussed in more detail in Chapter 3) included debates about the place of God (see discussion in Barnett 2003). Feminist perspectives, such as those introduced by Mary Wollstonecraft (1989 [1787–1797]), also introduced internal challenges to Western Enlightenment thinking – a challenging which continues today as, for example, in the feminist ethics of care explored in Chapter 3. Western philosophical traditions also encountered other traditions such as Islamic mathematical and scientific ideas, particularly in the Medieval period. However, as Iqbal (2000) points out, the influence of these ideas was (largely) elided in the Western Enlightenment and scientific and mathematical thinking in the seventeenth and eighteenth century. Hebraic and Hellenic (western) cultures and ideas have also encountered each other. For example this is reflected in the thinking of Philo in the first-century CE; the seventeenth-century thinker Spinoza, who developed ideas of God and nature which sat outside of both Christianity and Judaism (ideas being drawn into contemporary thinking as I explore in later chapters); and more recently in the work of thinkers such as Emmanuel Levinas, Martin Buber and Hannah Arendt.

Whilst acknowledging these internal and external challenges and developments within Western thinking, the ideas of Pythagoras and Plato continue to influence strongly present day Western (Eurocentric) mathematical, scientific and philosophical ideas. Sacks (2002: 48) calls this continuing influence of particular Greek roots 'Plato's ghost', which 'has haunted the Western imagination' for over 2500 years. Sacks uses Raphael's painting *The School of Athens* to illuminate Plato's thinking. This painting shows the grand, classical arches and columns of the school populated by groups of philosophers discussing, gesticulating, disputing. At the centre of the picture stand two figures: an older man, with a white beard, pointing upwards – Plato, and by his side, a younger man, his pupil – Aristotle. Sacks (2002: 49) suggests that Plato is saying that if Aristotle wishes to seek 'truth' he should not look for it down here in the messy and chaotic particulars of the world. Instead he should look 'heavenwards' – outside of this world. As Plato argued in *The Republic* (2007 [380 BCE]), using his parable of the cave, the world we see here is a 'mere play of shadows'. Outside this realm of messiness there exists harmony and unity of form beyond the here and now: a place where the true essence of a thing exists. It is there that 'trees become "*treeness*", men become "Man" and apparent truths coalesce into "Truth"' (Sacks 2002: 49) – a world of order set against the chaos of life.

These dominant Western (Eurocentric) framings of the world are problematic in a number of ways, including in education. First, Plato's emphasis that there are singular essences of things creates a universal truth about that thing which is then true for everyone for all times. This denies a place for particularity and difference and provides opportunity for those with power to impose their truth as universal. Moreover, such 'truths' are often so engrained in the thinking of the dominant group they are not even recognised by those who hold them. Other ways of knowing or being are ignored or denigrated as 'primitive'. Universalism is a barrier to the sustainable and democratic education called for in this book since it blocks the recognition of the emergence of other ways to be and be together in the world we share.

Second, the emphasis in the thought of Parmenides and Plato that forms are ultimately fixed encouraged a line of thinking in which the world can be broken down into stable, separate parts. This line of thought strongly influenced early European Enlightenment philosophical, mathematical and scientific ideas in the seventeenth century. It presented the attractive possibility that ultimately order is dominant in a world of seeming chaos, introducing mechanical understandings of the world. Examples of such mechanistic thinkers include the philosopher and mathematician René Descartes whose work 'offered a new vision of the natural world that continues to shape our thought today: a world of matter possessing a few fundamental properties and interacting according to a few universal laws' (Hatfield 2018: 1); the mathematician and scientist Isaac Newton and the chemist Robert Boyle. Broadly speaking, in this line of thinking, the world is broken down into parts which can be moved around but remain themselves unchanged. The world is a mechanical or clockwork process with 'man' (moreover this is a particular version of 'man' as I explore in Chapter 4) as the master of the mechanism. A system, irrespective of the number of changes it undergoes, can be reversed and return to its original ('fixed') state. However, as complexity thinking (which is discussed in more detail later in the chapter) emphasises, whilst mechanistic understandings might work in mechanically organised systems they are problematic for many areas of the observable world. Here systems interact with their environment, thus allowing for the possibility of emergence of the novelty necessary for sustainability understood in the ecosystem terms I outlined in the introductory chapter. The technologies developed in the Western Industrial Revolution, as well as the concurrent development of Modernism (which deliberately rejected ideas of the past and emphasised rationality, innovation and scientific developments), reinforced mechanistic thinking and understood nature as 'brute and inert'.

Third, framing nature as 'brute and inert', part of a clockwork mechanism with man in charge of the mechanism, encourages an approach where nature is understood as a resource to be used by humans rather than as something with a liveliness and value independent of human use. As I discuss in more detail in Chapter 8, the technologies developed in the Western Industrial Revolution coupled with these conceptions of nature as a resource for human use enabled capitalism to develop as a powerful force. In addition, demand for the materials needed by capitalism was a key driver in European colonialism in which European countries sought to dominate and exploit other countries – their people and wider nature – to supply industrial processes. It is beyond the scope of this book to explore colonialism, but it had the effect of exploiting countries, dehumanising Indigenous peoples and also spreading around the world particular Western religious, cultural and philosophical ideas which are still present today, including in education.

Whilst Parmenides and his inheritors emphasise thinking in which the world is framed as stable and nature as brute and inert, another Ancient Greek thinker, Heraclitus (2003 [500 BCE]) proposed a different way to understand the world. For him it was as a place of flux and emergent change rather than a place of underlying stability somehow outside time and place. His framing, however, did not go on to be as influential on the development of dominant Western thinking as those of Parmenides and Plato. Heraclitus drew from, but also stood in contrast to, his neighbours and predecessors the Milesian material monist philosophers such as Thales, Anaximander and Anaximenes (see Graham 2015, Cohen 2016). The Milesians proposed that the world was formed from some universal fixed, unchanging 'stuff' such as air or water that is the basis of other things. The Milesians understood change as variations of this basic, underlying, unchanging, material 'stuff'. Heraclitus (2003 [500 BCE]) reversed this: *change is what is real*. What is experienced as permanence is only apparent, a pattern in a world which is constantly in flux. His most famous example of flux and patterns, cited at the start of the chapter, is loosely translated as 'You never step in the same river twice'. The river is a pattern, but the water within it is in constant movement and change.

In his work, Heraclitus (2003 [500 BCE]) uses the *metaphor* of fire (he was not suggesting that the world was composed of fire) to represent change or flux, stating:

> That which always was,
> and is, and will be ever living fire,
> the same for all, the cosmos, made neither by god nor man
> replenishes in measure
> as it burns away. (Fragment 20)

However, change was not the only important feature of Heraclitus' thinking. For him, patterns are 'held together' through constant tension between opposites. In one sense there is the succession of the opposites, or contraries, such as night following day, death following life:

> By cosmic rule,
> as day yields night,
> so winter summer,
> war peace, plenty famine.
> All things change.
> Fire penetrates the lump of myrrh, until the joining
> bodies die and rise again
> in smoke called incense. (Fragment 36)

There is also a sense in which Heraclitus' idea of a unity of opposites involved more than just the succession of opposed states that occurs in cases of change. He suggests that:

> The cosmos works by harmony of tensions, like the lyre and bow.
> (Fragment 56)

The bow appears to us to be static, but it is in fact in dynamical relationship with an equal and opposite force exerted by the arms of the instrument or the archer. These opposing forces, held in (con)tension (ἡ ἔρις), create dynamic cohesion.[1] To cite Heraclitus:

> From the strains of binding opposites comes harmony.
> (Fragment 46)

I acknowledge the viewpoint proposed by thinkers such as Crutchfield (2002, 2011) and Pueyo (2014) that there is a tendency for natural systems to balance order and chaos (BOC). However, drawing on this Heraclitean line of thinking, if these equal and opposing forces were, for some reason, interrupted this can allow a system's underlying instability and tendency towards flux to emerge. This opens the possibility for new structures and organisations to appear.

Viewpoints drawing on a Heraclitean understanding of a world made of change/flux held in patterns through a process of *contention (ἡ ἔρις) are now coming to the fore* in Western science, education, business and philosophy (for example see Hawkins and Schmidt 2010, Müller-Merbach 2006, O'Connell 2006, Osberg 2015). The philosopher Bergson made an important contribution to this shift in thinking at the start of the twentieth century. Central to Bergson's philosophy is his concept of *élan vital* 'the explosive force – due to an unstable balance of tendencies – which life bears within itself (1912: 103). For Bergson, *élan vital* provides a way to understand the 'driver' for the *emergent change* which is a key feature of life itself and the basis of the 'fundamental irreducibility of life to matter' (Vaughan 2007: 16). *Élan vital* is not some kind of separate force or spirit planted inside the materiality of things in the world – something distinct from physiochemical forces which cannot be understood mechanistically. For Bergson, *élan vital* is an intrinsic part of something's materiality, but it is something *irreducible to mechanistic understandings*. As Vaughan (2007: 16–17) comments:

> *Elan vital*, then, does not simply signify a force different in kind to material forces. It signifies a force different in kind to *matter conceived mechanistically* (original italics) and this force is nothing more than that very same matter conceived intuitively: as active, creative, as itself vital – the very qualities that a mechanistic materialism effaces when it isolates superposable parts and treats as quantifiable and repeatable what is continuous qualitative change.
>
> (16–17)

Central to Bergson's idea of time is the notion of *irreversibility*. Time, for Bergson, is *not* understood as a separate series of moments, as in the separate frames of a film which can be run forward or back, each one a blank waiting to be inscribed. Bergson's notion of time is one of *irreversible time*, characterised by a continuous flow or duration. He calls this heterogeneous time and comments that (1912: 4–5):

> There is no stuff more resistant or substantial [*than time*]. For our duration is not merely one instant replacing another; if it were there would never be anything but the present – no prolonging of the past into the actual, no evolution, no concrete duration. Duration is the continuous progress of the past which gnaws into the future and which swells as it advances. And as the past grows without ceasing, so there is no limit to its preservation. Memory as we have tried to prove[2] is not a faculty of putting away recollections in a drawer or inscribing them in a register. There is no register, no drawer. . . . In reality the past is preserved automatically by itself.

This understanding of time introduces the notion of **material historicity** into the life of a system over time. Osberg (2015: 29–30) explains that material historicity 'relates to the fact that emergent processes are irreversible because their history of change operates as *a material structural element* (my italics) that acts together with the structural elements to play a part in determining the system's future'. Thus, 'the system's history can be understood to have been 'written into' or embodied in the system's physical structure' (for example see discussion in Holland 1998, Prigogine and Stengers 1984).

The issue of novelty, the existence of something radically new (emergents) which did not exist before, and could not be predicted even after the event, was highly problematic for Western science in the twentieth century. Such 'vitalistic' conceptions ran counter to an increasingly deterministic view of science dominated by causal (mechanical) relationships and the ideas of Bergson fell out of favour. More recently, framing the world as a place of flux and emergent change rather than a place of underlying stability, somehow outside of time and place (as in Plato's metaphysics/philosophy), has found expression in the area of complexity theory and emergence. This is the focus of the next section, in particular in relation to what these ideas can contribute to education, especially to education engaging with sustainability.

Flux and emergent change: ideas from complexity theory and the possibility of sustainable education

Complexity theory highlights how whilst mechanistic understandings might work in mechanically organised systems, such thinking is problematic for many areas of the observable world. The world is not a clockwork mechanism. In the natural world 'open systems do interact with their environment, changing themselves and their environment in the process' (Osberg and Biesta 2007: 49), allowing for the possibility of sustainability – for the emergence necessary for ecosystems to survive and develop.

In complexity thinking the possibility of emergence and novelty occurs when:

> very large numbers of constituent elements or agents are connected to and interacting with each other in many different ways.
>
> (Mason 2009: 119)

Such interactions:

> cause organisation and reorganisation and, if sufficient or critical level complex interactions are allowed to develop, new and sometimes surprising patterns and structures can emerge which are more than the sum of their parts.
>
> (119)

At this point it is important to note that ideas of complexity theory and emergence are not homogeneous (see Alhadeff-Jones 2008, Horn 2008, Osberg

2015). In the fields of non-linear mathematics and complexity science, an understanding of emergence has been developed which can be reconciled with deterministic (cause and effect) logic. Novelty does occur, and this cannot be predicted in practice in advance. However, once the novelty has emerged it can then be traced to the logical, deterministic rules which have created the novelty (Holland 1998). The system would reproduce the same novelty if the system were 'run' again. Thus, at least theoretically, the novelty is predictable *a priori*. It is this understanding of emergence which Osberg (2005) and Osberg and Biesta (2007) (drawing on Chalmers 2002) call 'weak' emergence.

A different understanding of emergence is provided by the chemist Ilya Prigogine. He was awarded the Nobel Prize in chemistry in 1977 for his work researching novelty and emergence in 'far from equilibrium' open systems. These are systems open to influences or perturbations outside the system and where inputs and outputs of energy and matter are not evenly balanced. These perturbations create turbulence which shifts the system far from its existing equilibrium point. When sufficient turbulence is reached a process of self-organisation at the micro level occurs which creates 'a process whereby properties that have never existed before and, more importantly, are inconceivable from what has come before, are *created* or somehow *come into being* for the first time' (Osberg and Biesta 2007: 33). The system 'jumps' to a new level of organisation, a new equilibrium. Prigogine calls the new properties in the system dissipative structures. A key feature of this process is what Prigogine (1977, 1983) calls 'points of undecidability' or 'bifurcation points' (see Figure 2.1). These are points where the system has to go one of several ways, and it is impossible to determine in advance what this will be. Repeating the experiment would not necessarily give rise to the same outcome.

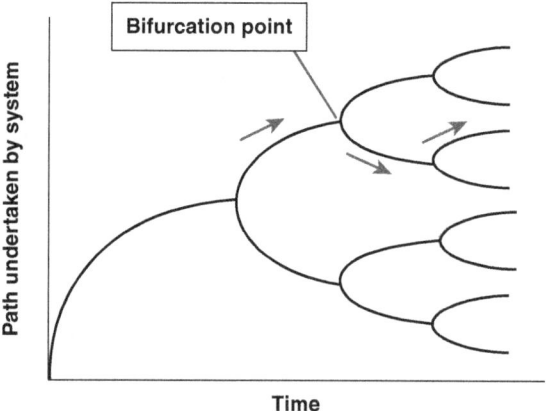

Figure 2.1 The development of bifurcations in far from equilibrium systems showing the unique path followed in an open system

Dissipative structures (the term Prigogine uses) or emergents (the term widely used in complexity thinking) cannot be predicted even in theory *a priori*. It is this understanding which Chalmers (2002), Osberg (2005) and Osberg and Biesta (2007) call 'strong' emergence.

Prigogine acknowledges the influence of the thinking of Bergson, particularly ideas expressed in *Creative Evolution* (1912). Osberg (2015) draws our attention to Prigogine's Nobel Prize acceptance speech in which he said:

> Since my adolescence I have read many philosophical texts, and I still remember the spell *L'Evolution Créatrice* cast on me. More specifically I felt some essential message was embedded, still to be made explicit, in Bergson's remark 'The more deeply we study the nature of time the better we understand that duration means invention, creation of forms, continuous elaboration of the absolutely new'.
>
> (Prigogine 1977: online)

There are a number of implications for education informed by the logic of complexity theory and emergence. First, understandings of what constitutes knowledge are challenged. If one accepts a line of argument in which education is, or has the potential to be, highly complex, messy and unpredictable (the conditions which make emergence a possibility), operating in what could be called strongly emergent open systems 'Knowledge is neither representation of something more "real" than itself, nor an object that can be transferred from one place to the next. Knowledge is understood, rather, to emerge as we human beings participate in the world' (Osberg and Biesta 2008: 313). Furthermore, 'every meaning that emerges is uniquely new, something which has not been in the world before, and cannot be predicted from the ground from which it emerged' (Osberg and Biesta 2008: 313).[3] It then follows that if knowledge and education are seen in this way it is necessary *to identify and organise opportunities for this kind of emergent learning to take place*. Education is no longer about moving knowledge from one place to another (usually from the teacher to the student). Rather, it is about creating a setting where *learners are active participants in the making of emergent meaning*.

Second, complexity thinking opens possible alternatives to a 'curriculum for survival' (Curren 2009) in this era of the Anthropocene and climate change. A logic informed by complexity theory opens up emergence and potential for 'enlarging the space of the possible' – a *curriculum for potentiality*. Osberg (2015) identifies different approaches to arguments for adopting such a logic of potentiality within education. She highlights how some theorists engage with Deleuzian[4] approaches. Some of these 'Deleuzians' emphasise *immanence*. This draws on Bergson's theme of the 'uncontrollable yet profoundly life-generating nature of "irreversible becoming"' (Osberg 2015: 35). In such a framing:

> education and learning is conceptualised as becoming and the pedagogical task as a 'pluri-vocal' endeavour in which 'explanation' is replaced by

'presentation of ideas' as the pedagogue works 'to intensify connections in actuality while seeking to release new potentiality' (Semetsky and Masny 2013: 16).

(Osberg 2015: 35)

Braidotti (whose thinking I explore in Chapter 4) identifies her thinking as 'a branch of complexity theory' (Braidotti 2013:189) in this Deleuzian trajectory, emphasising immanence and 'enfleshed materialism' (Braidotti 2002). She thus forms a bridge connecting aspects of complexity thinking to recent new materialist debates (for example see Snaza *et al.* 2016 for discussion of new materialism and education). Braidotti emphasises all matter as vibrant and the traversing of boundaries between self/other in a process of becoming. Braidotti's understanding of *zoë* is a restatement of Bergson's (1912: 103) conception of *élan vital* – namely the vital force 'contained within life itself'. Other Deleuzian thinkers indicate that the emphasis on potentiality in the Deleuzian trajectory introduces a 'time dimension of later when learning is put into practice' (Semetsky and Masny 2013: 16). In such a framing, 'a genuine education is characterised by a transcendent movement of change, as much as being immanent to experience and events' (Semetsky and Masny 2013: 16 cited in Osberg 2015: 35).

Another complexity-compatible approach to theorising curricula of potentiality follows a Derridian line of thinking (see for example Osberg 2005, 2010, Osberg and Biesta 2007, 2008). This trajectory builds on Prigogine's notions of 'undecidability' and bifurcation.[5] In such a viewpoint the emphasis is on 'the moment in which the choice [of the system] is undetermined, the moment of free play' (Osberg 2015: 35). It 'is the moment in which agency becomes possible' (35). Since the moment is 'undetermined', it is not being informed or constrained by notions in the present such as survival. The choices made transcend and are superfluous to the present, and ideas in the present, thus making thinking 'the unthinkable' possible. Osberg suggests that whilst learning about survival *is* important, and has an important role within education, it is not 'all there is to learning' (36). Learning can also be about 'the free play of ideas . . . an experimental attitude in which multiple possibilities present themselves *as equal possibilities* with no predetermined 'best' solution (even in retrospect) (36)'. The future is understood to be 'radically open ended, filled with creative and as yet unimagined potential' (36), an 'understanding in which the truly creative – that which serves no immediate function, but for which a new function will ultimately be found- still has a place in our understandings of what it is to be human' (36).

Osberg's ideas on a curriculum of potentiality resonate with Facer's (2016: 58–59) conception of a 'pedagogy of the present' in which the:

educational encounter is understood as constituting a distinctive temporality of its own, a temporality that is characterised by the intentional putting into play of the abundant materials that constitute and create futures, that keeps them intentionally open for exploration, and that expands and encourages the space to participate in that process.

The theoretical insights argued for in this book draw on both immanent and transcendent understandings of emergence and potentiality and what such understandings can offer to education as both democratic and sustainable emergent processes in a 'pedagogy of the present'. Such a hybridisation is not a weakness or confusion of ideas. Rather, such a bringing together of different ideas opens the potential for new ideas to emerge.

In drawing together these ideas to propose education as having potential to be a highly complex, messy and unpredictable process, operating in strongly emergent open systems creating conditions necessary for complexity and emergence, some key ideas emerge. First, in addition to new understandings of the role of the teacher in relation to knowledge, *the issue of subjectivity – what it is to be a human subject who acts in the world – also needs to be left radically open.* Subjectivity too can no longer be assumed in advance, opening possibilities to explore lines of thinking compatible with both post-humanist and posthuman thinking, as I do in subsequent chapters. Second, the *relationship* between the teacher and student changes. The teacher's role is no longer (only) to transmit knowledge but instead to create conditions where 'critical level complex interactions are allowed to develop' (Mason 2009: 119); where meaning can emerge; where there are opportunities *for playful exploration and the creation of a surplus.* Such emergent processes have the potential to encourage ideas which can include, but also *go beyond,* the constraints of the needs of today, opening as yet unknown possibilities and uses in the future. The nature of the relationship opens opportunities for teachers and learners to shape each other's subjectivities in an inter- and intra- active 'dance of encounters': a process of 'making-with' which Haraway (2016: 58) calls 'sympoiesis'.

Whilst exploring the positive aspects of drawing on complexity and complexity-compatible thinking in education, it is also important to review and address some of the objections to drawing on a logic of complexity in education. First, as the discussions of strong and weak emergence demonstrate, complexity theory has many facets and trajectories. These draw on different philosophical and scientific traditions with a potential for conflict to occur between proponents of these differing views. However, this is the time to be aware of what Latour (2004) calls the need for theorists to gather around areas of concern and Haraway (2007) calls 'matters of care' to see what each perspective can contribute to the issue rather than argue amongst themselves about the different points of view. Indeed, when different perspectives are brought together, whether these are differing ideas within a particular discipline or perspectives from different disciplines such as science, arts and philosophy, there is the possibility of emergence of the new. This is especially so if the ideas are discussed on the premise that all participants approach such activities on an *equal* basis. The CERN Institute,[6] recognises the potential of such an approach in its

Artists-in-Residence Programme Collide@CERN. Its director Arianne Koek, cited in a review by Kieniewicz (2013: online) comments that:

> The space of collaborations between artists and scientists is a 'troubled' area, where artists and scientists rarely stand on an equal footing as professionals. In an art/science collaboration, what is the role of the artist in relationship to the scientist? A basic premise of the Collide@Cern programme is to place artists and scientists on the same level.

The programme expects excellence from artists and scientists. It does not demand from artists that their work should only communicate the work of the scientists at the Institute to the wider public. Their role is also to be part of a wider creative endeavour, including assisting the scientists to look at and understand the world differently. Koek comments that to achieve this it is not enough 'to simply put an artist in the lab with a physicist and expect them to figure out how to play nicely'. Koek explains her role 'as a kind of producer-in-residence, working with the artist-in-residence and CERN's scientific community to facilitate 'creative collisions'. This example is *not* provided as a 'blueprint' for action. Rather, it is an example of the possibility of working in unusual ways to explore areas of concern and *also* the possibilities that such approaches can open up. An example of such activity organised by the artist-in-residence was a visit to some newly discovered tunnels/caves in the CERN complex. The artist-in-residence and scientists explored Plato's conception of the cave, introduced earlier in this chapter, in which what one sees are mere shadows of ideal stable forms. They explored how this still influences Western conceptualisation of the world as essentially static and stable, how this might limit scientific research and why this can be problematic.

Second, the focus on emergence of the new raises another important issue with using complexity frameworks within education – the issue of ethics. If the 'radically new' is allowed to emerge from a 'free play of ideas', this raises the question of what can or should be done if the 'new' is deemed 'unethical' and who gets to decide. This important and challenging issue of ethics is explored in Chapter 6.

Third, some theorists (for example see Hunter 1997, Hunter and Benson 1997) argue that complexity frameworks do not add ways to understand education which are not already sufficiently explained elsewhere. They refer to recent developments in post-structural and critical theory, the ideas present in the work of Piaget and Dewey and Whitehead's organic process philosophy. They argue it is unnecessary to have new theories for theory's sake. However, such a viewpoint can be countered in several ways. Kuhn (2005, 2008) suggests that such thinking is unhelpful and evidence of 'boundary keeping': protecting one's own research interests in an academic environment which is set up in increasingly competitive ways. Such boundary keeping closes down rather than opens up new ways of thinking and knowing. Osberg (2015: 36) argues that speaking of complexity and complexity-compatible theoretical approaches as a replacement to existing frameworks is to conceive of such approaches in too narrow a sense. She suggests that:

> In referring to a 'complex' approach to learning one is not referring to a set of ideas that have been unified by complexity theory and which can, as some fear, be used indiscriminately to colonise existing theorisations of learning but to an ongoing and continuously hybridising movement.

Indeed, she argues that it is this hybridising of ideas which gives them much of their vigour. Hybridisation is at work in approaches discussed earlier in this chapter of both immanent and transcendent conceptions of emergence, drawing as they do on Derridian and Deleuzian perspectives. It can also be seen in theorising which re-examines existing theoretical work through the lens of a complexity theory. In the 2008 special issue of *Educational Philosophy and Theory*, which focused on complexity theory and the philosophy of education, a range of theorists and their ideas are explored and reinterpreted through the lens of complexity thinking. These include Tyler and Vygotsky (Doll 2008), Foucault (Olssen 2008) and Dewey (Semetsky 2008). Further examples of exploring thinkers through the lens of complexity theory include Osberg and Biesta's examination of Derridian notions of representation (Osberg 2005), Osberg and Biesta (2007, 2008) and Doll's (1993) discussion of new ways to conceptualise the curriculum using ideas drawn from Piaget and Dewey. In these examples, logic drawing on complexity theory is employed to '*perturb*' (Davis and Sumara 2006) not to replace existing theorisations.

A fourth possible issue when proposing the use of complexity and complexity-compatible frameworks in education relates to the question of whether concepts from science and mathematics can so easily be transferred to the social sciences, including education. For example, Hunter and Benson (1997) argue that complexity theory has too deterministic an understanding to be valuable for exploring social interactions. They argue that suggesting that such deterministic scientific theories can explain and contribute to the educational process is misplaced, citing behavioural models as an example of another such flawed endeavour. A review of their approach indicates that they are drawing on models of 'weak' emergence such as that of Holland (1998), which do indeed have such a deterministic orientation. 'Strong' understandings of emergence, as discussed earlier, do *not* have this deterministic aspect. Indeed, towards the end of their own article, Hunter and Benson (1997) comment that the then more recent understandings of complexity such as that of Doll (1993) do demonstrate features of indeterminacy. It is also possible to question the assumption that complexity thinking has simply been 'transferred' from science and mathematics to the social sciences. Kuhn (2008: 184) argues that to think that ideas develop in compartmentalised disciplines is too narrow. He proposes that whilst some of the language used in complexity thinking may be borrowed from science and mathematics, new thinking which questions existing epistemological assumptions can arise simultaneously across many disciplines, each contributing and helping to shape the others. For example, he suggests that it could have been social science's explorations of the complexity and indeterminacy of human behaviour which encouraged scientists to question their own understandings of determinacy and linear causality.

Finally, if we do accept that systems thinking and concepts of complexity and emergence are not just being used in education as a 'metaphor', this raises the question of which part of the educational process is being referred to: the societal or school level, the level of the classroom as a complex system or the learning process in the brain of each individual.[7] If all of these are examples of complex systems this then raises the question of how these different understandings fit together. Hetherington (2012) suggests that one way this difficulty has been addressed in the literature is by the concept of nested complex systems (Davis and Sumara 2006). Smaller systems can be nested within a larger system. It is worth noting here, however, that understandings of 'nested systems' emphasise the *spatial continuity* of systems with an underlying conception of an inherently stable whole. In this book the focus is on the logic of *flux* and *temporal continuity* within a system – an extension to current ideas in systems thinking, not a replacement of them.

The various ideas and understandings of a world in flux explored thus far in this chapter can be understood as contributing to opening spaces, including opening spaces in education, which can enlarge possibilities in and for the world we share. I have traced ideas from Heraclitus through Bergson to complexity theory and *strong emergence* as a way to challenge static conceptions of the world; the 'ghost' of Plato and his predecessor Parmenides and mechanistic understandings of the world in which 'nature' is understood as 'brute and inert'. It is important, however, to recognise that alternative framings of the world developed *within* Western (Eurocentric) thinking are not the only ones which call into question static/stable framings of the world. It is to this issue I now turn.

Indigenous thinking and a world in flux

Framings which understand the world as a place of flux have always been a part of the thinking of many Indigenous peoples but have not received recognition or been valued in a world dominated by particular universalised framings. However, as I highlighted in the introductory chapter, if one wants to engage with these Indigenous ways of knowing and being in the world, it is very important to both be aware of and try to minimise the risks of appropriation – a 'using of' and 'benefiting from' such ideas rather than 'engaging with' and 'learning from' them. It is also important to recognise the harm which has been caused and which continues to be caused to Indigenous people by dominant

Western (Eurocentric) power structures and attitudes. The suppression and denigration of Indigenous ideas and ways of being has led/leads to an 'other-ing' and ultimately to what Jackson (2013) calls the 'unhumaning' of Black and Indigenous peoples in Eurocentric framings. One cannot 'borrow' from Indigenous ideas at the convenience of Western scholarship without also work-ing to identify and interrupt continuing harmful treatment (see discussion in Andreotti 2014, 2016, Todd 2016). It is also important to be aware of the dan-ger of 'grafting' Indigenous knowledges into other settings without grasping their full ontological significance (see Ahenakew 2016). Battiste (2011 [2000]) argues that it is vital that Indigenous knowledge be defined and contextualised by Indigenous people.

Faced with such issues it can be challenging to find a way forward if one wants to engage with Indigenous knowledge without appropriation. Todd (2016) suggests a good starting point is to cite Indigenous scholars directly rather than using literature written by Western scholars for whom Indigenous people and thinking are 'objects' of study. Todd also recommends Sundberg's (2014) toolkit as a method for approaching Indigenous knowledge in ways which can reduce appropriation.

Sundberg highlights two 'performances' which commonly affect engage-ments between Eurocentric and Indigenous knowledges. The first is an omission of one's *loci of enunciation* – the places one is speaking from. Not acknowledging that one is speaking from a particular viewpoint then runs the risk that ideas generated are shared as being 'universal' – applicable to all, and also new for all. The second performance relates to the particular ways that Indigenous knowledges are summoned or spoken about. These can be portrayed as magical, primitive, methodologically naïve – somehow 'other' – viewed from a position of ontological superiority. To respond to these dam-aging performances Sundberg suggests a toolkit drawing on Spivak's (1993) 'homeworking' – the 'locating our body-knowledge in relation to the existing paths we know and walk' (Sundberg 2014: 39) and the Zapatista movement's *preguntando caminamos*.

The first step – 'homeworking' is a 'self-reflexive activity of becoming aware of one's own ontological and epistemological assumptions' (39). It requires engagement with how these positions have become naturalised in and through *power relations*; the *privileges* that these have afforded; *harms* these have caused/continue to cause and the *white discomfort* which facing these issues generates. 'Homeworking' is central to the work this book aims to do.

The next step, the Zapatista movement's *preguntando caminamos* is an invita-tion to 'walking with' and to 'asking as we walk'. 'Walking with' has at least two meanings. In one sense 'walking with' means 'having a mutual respect for a multiplicity of life worlds' (Zapatistas Army of National Liberation 2005 cited in Sundberg 2014: 40). It is also about how to learn about multiplicity. This requires a shift from a 'will to know' towards learning as 'an engagement with' which may also entail 'a learning from'. In this book I aim to create an 'atmosphere of 'engagement with' and 'learning from' by careful attention to ideas shared by Indigenous thinkers. This includes signposts to the writings and also other forms of expression of Indigenous scholars for whom writing is often not the first choice.[8] I introduce longer direct quotations rather than paraphrasing when trying to share ideas of Indigenous thinkers, whilst at the same time recognising that I have selected these quotations with the potential for bias and also 'grafting' this can bring. I acknowledge that these actions are not 'enough', but I hope I demonstrate my awareness of issues and the need for 'walking with' and 'learning from'.

A well-known Indigenous academic who has written about Indigenous thinking/science on a world in flux is Leroy Little Bear. He was the founder of the Native American Studies Department at the University of Lethbridge and also the founding Director of Harvard University's Native American Program. He is a member of the Small Robes band of the Blood Indian Tribe of the Blackfoot Confederacy (Native American Academy, online). Through his learning/exposure to Indigenous and Western ways of thinking and being, he can explore both systems, as well as the boundaries and 'borderlands' between them.[9] He acknowledges that his Indigenous knowledge is based on the philosophy of the Plains Indians but highlights how 'there is enough similarity among North American Indian philosophies and beyond to apply the concepts generally, even though there may be individual differences or differing emphases' (Little Bear 2000: 77).

Little Bear (2011: online) identifies three tenets in (his) Indigenous thinking, commenting:

> The first tenet of the Native paradigm is what we refer to as constant flux. If you were to imagine this flux is animated, you would see a constant motion or energy waves, light and so on, going back and forth. Things are forever in motion, things are forever changing. There is nothing certain. The only thing that is certain is change. Things are forever moving, things are forever dissolving, reforming, transforming.
>
> A second part of the native tenet of flux is flux itself. Everything in existence, everything in creation, consists of energy waves. In classical [*Western-* my addition] physics, we talk in terms of matter, particles, subatomic particles. In the native way, we talk in terms of energy waves. Those energy waves are very special because it's those energy waves, not you, that know. All of us are simply combinations of energy waves. Spirit is energy waves. All it means when we die is that particular combination becomes dissipated. Energy waves are still there.
>
> A third part of the paradigm is that everything is animate. There is nothing in Blackfoot for instance, that is inanimate. Everything is animate. Everything, those rocks, those trees, those animals all have spirit just like we do as humans. If they all have spirit, that's what we refer to as, 'all my relations'.

He highlights how the idea of all things being in constant motion or flux leads to a worldview which is both holistic and cyclical, commenting:

> If everything is constantly moving and changing, then one has to look at the whole to begin to see patterns. For instance, the cosmic cycles are in constant motion, but they have regular patterns that result in recurrences such as the seasons of the year, the migration of the animals, renewal ceremonies, songs, and stories. Constant motion, as manifested in cyclical or repetitive patterns, emphasises process as opposed to product. It results in a concept of time that is dynamic but without motion. Time is part of the constant flux but goes nowhere. Time just is.
>
> (Little Bear 2000: 77)

There is a difference here from complexity thinking in which time is unidirectional (towards the future). When working at the borderlands between different knowledges, it is important to acknowledge the incommensurabilities in different ways of understanding the world and to explore what these can mean for the ways we are together in the world rather than try to find an agreed 'one way', which then suppresses other possibilities. I explore this issue of living with such difference in more detail through consideration of the thinking of Chantal Mouffe on agonistic pluralism in Chapter 5.

Little Bear highlights that land is important in Indigenous thinking since land, whilst still also in flux, gives us some sense of stability and takes a central position in learning and ways to be together in the world. He contrasts this with Western thinking in which time has a stable, central position. In such Western thinking one moves through different spaces/places, but time is a constant that one can take, compare, rationalise. Think of sets of clocks in stock exchanges which compare the time in different time zones, e.g. London, Paris, New York, Sydney. Teila Watson,[10] a Birri Gubba and Kungalu Murri woman, highlights that a common theme across the thinking of many different Indigenous peoples is that 'land is held sacred in the production of life' and 'this relationship between people and land lays the foundation for the relationship between people' (Watson 2017: 1). Watson emphasises that this is not to present an overly romanticised view of Indigenous thinking and ways of being, acknowledging tough times, difficulties and conflict between people, but 'the conflict and toughness was mitigated by our philosophy, logic, ontology and law' (1).

Engaging with Indigenous knowledge on flux can enable a challenging of dominant Western ontologies, an opening up of 'learning from' and possibilities of new ways of thinking and being which draw in knowledge and understanding from different places, different wisdoms. This is not an easy process but nevertheless a worthwhile undertaking which this book hopes to facilitate. The Métis scholar and artist De Line (2016) engages with the processes and significances of bringing together different ways of knowing and being (epistemologies and ontologies). In his writing he brings the Indigenous science on *Niw_hk_ m_kanak* (all my/our relations) into conversation with Western quantum physics, in particular the work of Karen Barad on intra-action. De Line highlights how *the words Niw_hk_ m_kanak* (all my/our relations) are spoken during the opening and closing ceremonies used by Cree and Métis Nations including Blackfoot and Haudenosaunee Confederacies. These words:

> acknowledge and bless all that is in the continuum, continually in flux, in all our relations: all energy waves that are in contingent relationality through a familial network. *All my/our relations* is not only a ceremonial acknowledgement but a philosophical proposition.
>
> (2)

De Line emphasises the significance of 'familial relationship' in all/our relations. Family relations go beyond interconnectivity, introducing *responsibility* – an ethical duty of care for the other. Little Bear (2016) also emphasises the connections and family relations held between humans and the wider natural world, commenting on the ethical responsibility that such a web of relations places on us. He comments that when we understand the earth as our Mother, we then need to think about how we are treating our Mother.

In his bringing of Indigenous and Western science into conversation De Line emphasises that how language is used is important. Indigenous science is just as much science as Western science.[11] To explore this point he highlights the diffraction which occurs in Western Science's double-slit experiment used in quantum physics. In this experiment, light or water pass through a barrier with two openings, creating diffraction patterns. It is beyond the scope of this book to make a study of quantum physics, but De Line identifies that a key outcome of the experiment is that 'energy, on a quantum level, can behave unpredictably producing contingent outcomes: electrons deviate whilst being monitored by a recording device during the experiments, forming irregularities in their behavior' (2). De Line then proposes that if one accepts water is made of quantum matter, with its *unpredictability* demonstrated through the irregular deviations of its electrons, this opens ways for Western science to understand water as *animate*. Water is no longer understood as 'brute and inert'– raising amongst other things, philosophical and ethical questions about the rights of water and human ethical relationships with water. Animate framings challenge Western ways of identifying water as a resource to be used rather than engaging with water as an active participant in the web of relations in the world.

In exploring Karen Barad's thinking on agential realism in which she comments:

> Knowing, thinking, measuring theorising and observing are material practices of intra-acting within and as part of the world. . . . Interacting with

(or rather intra-acting 'with' and as part of the world) is part and parcel of seeing. Objects are not already there; they emerge through specific practices.

(Barad cited in De Line 2016: 6)

De Line highlights that it is the emergent properties of matter, the flux, which enables the radically new to appear. He does, however, takes issue with Barad's use of the word 'seeing': a word which he argues reveals a Western (supposedly objective) 'gaze' rather than an 'experiencing' which takes place within, and as part of, the web of all my/our relations. For De Line, bringing Western and Indigenous science into conversation is not to set up oppositions, nor is it an attempt 'to solve the problem of how we relate to each other' (De Line 2016: 6). Rather, the aim is to highlight the potential that bringing together these different ways of thinking and being, including commensurabilities and incommensurabilities, opens up. It is also an act of recognition of the *entangled* web of which these ideas are a part.

This issue of entanglement is central to the thinking of Dwayne Donald (2009), a descendant of the Papaschase Cree. Donald explores this issue of entanglement using the idea of 'forts' and 'camps outside the fort'. He recalls visiting a reconstruction of a fur trading fort in Alberta. Inside the fort was a re-creation of a nineteenth-century, white-settler, fur-trader post. Outside the fort was an 'Indian' encampment' with actors engaging in various 'Native American' activities including cooking, childcare and dancing: two cultures and knowledge systems kept apart, operating in different spheres. Donald argues that the situation with early trading was not in fact like this since historical records show interchange between the two knowledge systems and ways of life. He uses this starting point to argue for present day entanglement of ideas, including in school curricula where, he argues, a 'fort and camp' mentality persists. He emphasises however that such entanglement needs to be approached through what he calls a principle of 'ethical relationality':

An ecological understanding of human relationality that does not deny difference, but rather seeks to more deeply understand how our different histories and experiences position us in relation to each other. This form of relationality is ethical because it does not overlook or invisibilise the particular historical, cultural, and social contexts from which a particular person understands and experiences living in the world. It puts these considerations at the forefront of engagements across frontiers of difference.

(6)

This chapter began by exploring Western (Eurocentric) framings in which the world is understood as being ultimately stable and made of separate parts which can be moved and returned to their original position unchanged – a 'mechanistic worldview'. It also considered why this framing is problematic. Through engagement with ancient Western thinkers such as Heraclitus, twentieth-century Western thinking such as that of Bergson and complexity theories and also Indigenous knowledges and ways of being, the chapter introduced framings which engage with flux, emergence, entanglement and intra-action and ways in which all in the world are my/our relations. In the next chapter I examine ways in which dominant Western (Eurocentric) worldviews emphasise rationality and autonomy alongside their emphasis on a static world made up of separate parts. These explorations draw in the challenges, including ethical challenges, to rationality and autonomy created by framings emphasising flux, interconnection and relationality considered in this chapter.

Notes

1 See discussion in Osberg 2015, Robinson 1987.
2 Bergson includes a footnote here referring to his work *Matière et Memoire*, Paris 1896, Chapters 2 and 3.
3 In the introduction I used this quotation as a definition of the 'radically new', and in this book the word 'radical' is used in this sense.
4 Osberg uses the term 'complexity-compatible thinking' to describe theories and ideas which, whilst not identified as complexity theory, are compatible with it and sometimes share some of its historical sources. Such is the case with Deleuze who, through his book *Le Bergsonisme,* helped rekindled interest in Bergson within academia.
5 When a system has a number of choices in deciding which way to go, and this cannot be known in advance, nor repeated.
6 The European Organization for Nuclear Research and home to the Hadron Collider.
7 Education research literature provides examples of research at these different levels. For example, Hetherington (2012) used the primary science classroom as the focus for her research. Cunningham (2001) employs a complexity 'lens' to examine and critique research on school improvement at the whole school level.
8 For example, *Isuma* (Zacharias Kunuk, Norman Cohn, Paul Apak, Pauloosie Qulitalik) use film to capture the storytelling tradition of the Inuit people. Their online 'book', which is free to those with access to an internet connection, includes a growing archive which documents a practice that privileges media, material and narrative invention (Isuma 2019) over written text. Their approach highlights the opportunities that such media offer for expression of Indigenous traditions. Their film, *One Day in the life of Noah Piugattuk*, was the feature of the Canadian Pavilion at the Venice Biennale 2019.
9 The term borderlands/border thinking was 'first used by Gloria Anzaldúa (1999) in her book *Borderlands/La Frontera: The New Mestiza*' (globalsocialtheory.org: online). It was further developed by decolonial thinkers, most notably Walter Mignolo, who comments 'Border thinking is the epistemology of the exteriority: that is, of the outside created from the inside' (see Mignolo and Tlostanova 2006: 206). Border thinking emphasises that 'the theoretical and epistemic must have a lived dimension' (globalsocialtheory.org: online) and recognises that theories exist at the borders and outside the colonial matrix of power. 'Lived' here 'is in the sense of the experience of those who have been excluded from the production of knowledge by Modernity'. Border thinking does not happen irrespective of modernity but rather as a response

to it as part of real life struggles against the oppressive apparatus of the colonial mix of power' (globalsocialtheory.org: online).

10 Teila Watson is a singer, poet and lyricist known as an Ancestress, writer and actor. Her art practice is influenced by climate change, decolonisation, the impact of first nations knowledges and practice on country and people, and the importance of land rights and first nations sovereignty. She recently received The Australia Council for the Art's Dreaming Award.

11 Recognition of Indigenous science is noticeably absent in much academic work, for example in the writing of Pei *et al.* (2009) in their application of 'traditional knowledge' to forest management in South East Asia. Pei *et al.* value 'traditional knowledge and the contribution it makes to sustainable forest management', but, in their writing, it is only Western science which is referred to as 'science' (with the adjective 'Western' missing). Indigenous science is called traditional knowledge or referred to using various terms such as 'spiritual and cultural practices'.

References

Ahenakew, C. (2016) Grafting indigenous ways of knowing onto non-indigenous ways of being: The (underestimated) challenges of a decolonial imagination. *International Review of Qualitative Research*, 9(3): 323–340. doi: 10.1525/irqr.2016.9.3.323

Alhadeff-Jones, M. (2008) Three generations of complexity theories: Nuances and ambiguities. *Educational Philosophy and Theory*, 40(1), 66–82. doi: 2007.00411.x

Andreotti, V. (2014) Conflicting epistemic demands in poststructuralist and postcolonial engagements with questions of complicity in systemic harm. *Educational Studies*, 50(4): 378–397. doi: 10.1080/00131946.2014.924940

Anzaldúa, G.E. (1999) *Borderlands/La frontera: The new mestiza*. San Francisco: Aunt Lute Books.

Barnett, S. (2003) *The enlightenment and religion: The myths of modernity*. Manchester: Manchester University Press.

Battiste, M. (ed.) (2011 [2000]) *Reclaiming Indigenous voice and vision*. Vancouver, BC and Toronto: University of British Columbia Press.

Bergson, H. (1912) *Creative evolution*. Mitchell, A. (tr.). London: Palgrave Macmillan.

Braidotti, R. (2002) *Metamorphoses: Towards a materialist theory of becoming*. Oxford: Blackwell Publishing.

Braidotti, R. (2013) *The posthuman*. Boston, MA and Cambridge: Polity Press.

Chalmers, D.J. (2002) *Varieties of emergence*. Manuscript written for the Templeton Foundation workshop, Granada, August. http://consc.net/papers/granada.html [Accessed 20.6.2016].

Cohen, S.M. (2016) Heraclitus. *Washington.edu*. Available at: https://faculty.washington.edu/smcohen/320/heracli.htm [Accessed 12.1.2017].

Crutchfield, J. (2002) *What lies between order and chaos*. Santé Fe Institute. Available at: http://citeseerx.ist.psu.edu/viewdoc/download?doi=10.1.1.84.6812&rep=rep1&type=pdf [Accessed 9.3.2020].

Crutchfield, J. (2011) Between order and chaos. *Nature Physics*. Available at: http://csc.ucdavis.edu/~chaos/papers/Crutchfield.NaturePhysics2012.pdf [Accessed 9.3.2020].

Cunningham, R. (2001) *Chaos, complexity and the study of education communities*. Paper presented to the British Educational Research Association Annual Conference, University of Leeds, 13–15th September. Available at: www.leeds.ac.uk/educol/documents/00001895.doc [Accessed 13.2.2015].

Curren, R. (2009) *Education for sustainable development: A philosophical assessment*. London: PESGB, Impact Series. Republished on-line. Available at: http://onlinelibrary.wiley.com/doi/10.1111/imp.2009.2009.issue-18/issuetoc [Accessed 31.12.2016].

Davis, B. and Sumara, D. (2006) *Complexity and education: Enquiries into learning, teaching and research*. London: Routledge.

De Line, S. (2016) All my/our relations: Can posthumanism be decolonized? *Open! Platform for Art, Culture & the Public Domain*. Available at: www.onlineopen.org/all-my-our-relations [Accessed 23.1.2018].

Doll, W.E. Jr. (1993) *Post-modern perspectives on curriculum*. New York: Teachers College Press.

Doll, W.E. Jr. (2008) Complexity and the culture of curriculum. *Educational Philosophy and Theory*, 40(1): 190–212. doi: 10.1111/j.14695812.2007.00404.x

Donald, D. (2009) Forts, curriculum, and Indigenous Metissage: Imagining decolonisation of Aboriginal-Canadian relations in educational contexts. *First Nations Perspectives*, 2(1): 1–24.

Facer, K. (2016) Using the future in education: Creating space for openness, hope and novelty. In H.E. Lees and N. Noddings (eds.) *The Palgrave international handbook of alternative education*. London: Palgrave Macmillan, pp. 53–78.

Globalsocialtheory.org (online) *Border thinking*. Available at: https://globalsocialtheory.org/concepts/border-thinking/ [Accessed 11.9.2019].

Graeber, D. (2001) *Toward an anthropological theory of value: The false coin of our own dream*. New York: Palgrave Macmillan.

Graham, D.W. (2015) Heraclitus. In E.N. Zalta (ed.) *The Stanford encyclopedia of philosophy*, Fall 2015 Edition. Available at: http://plato.stanford.edu/archives/fall2015/entries/heraclitus/ [Accessed18.6.2016].

Haraway, D. (2007) *When species meet*. Minneapolis and London: University of Minnesota Press.

Haraway, D. (2016) *Staying with the trouble: Making kin in the Chthulucene*. Durham, NC: Duke University Press.

Hatfield, G. (2018) "René Descartes". In E. Zalta (ed.) *The Stanford encyclopedia of Philosophy*, Summer 2018 Edition. Available at: https://plato.stanford.edu/archives/sum2018/entries/descartes/ [Accessed 4.7.2020].

Hawkins, A. and Schmidt, H. (eds.) (2010) *Handbook of optofluidics*. Boca Raton, FL: CRC Press, Taylor and Francis.

Heraclitus of Ephesus (2003 [500 BCE]) *Fragments*. Haxton, B. (tr.). London: Penguin.

Hetherington, L. (2012) *Walking the line between structure and freedom: A case study of teachers' responses to curriculum change using complexity theory* (Unpublished PhD thesis). University of Exeter, Exeter. Available at: https://ore.exeter.ac.uk/repository/handle/10036/3868 [Accessed 6.11.2014].

Holland, J. (1998) *Emergence: From chaos to order*. Reading, MA: Addison-Wesley.

Horn, J. (2008) Human research and complexity theory. *Educational Philosophy and Theory*, 40(1): 130–143. doi: 10.1111/j.1469-5812.2007.00395.x

Hunter, J.W. (1997) Review: Democracy, chaos and the new school order by S.J. Maxcy. *The Journal of Educational Thought/Revue de la Pensée Educative*, 31(1): 90–93.

Hunter, J.W. and Benson, G.D. (1997) Arrows in time: The misapplication of chaos theory to education. *Journal of Curriculum Studies*, 29(1): 87–100.

Intergovernmental Panel on Climate Change (IPPC) (2014) *Fifth assessment report*. Available at: www.ipcc.ch/ [Accessed 16.6.2014].

Intergovernmental Panel on Climate Change (IPCC) (2018) *Special report on global warming of 1.5°C approved by governments – summary for policy makers.* Available at: www.ipcc.ch/sr15/ [Accessed 14.1.2020].

Intergovernmental Science-Policy Platform on Biodiversity and Ecosystem Services (IPBES) (2019) *Global assessment report on biodiversity and ecosystem services.* Available at: www.ipbes.net/global-assessment-report-biodiversity-ecosystem-services [Accessed 28.8.2019].

Iqbal, M. (2000) Islam and modern science: Formulating the questions. *Islamic Studies – Special Issue: Islam and Science*, 39(4): 517–570.

Isuma (2019) *The Isuma book.* Available at: www.isuma.tv/isuma-book [Accessed 12.9.2019].

Jackson, Z.I. (2013) Review: Animal: New directions in the theorization of race and posthumanism reviewed work(s): HumAnimal: Race, law, language by Kalpana Rahita Seshadri; The birth of a jungle: Animality in progressive-Era U.S. literature and culture by Michael Lundblad; Animacies: Biopolitics, racial mattering, and queer affect by Mel Y. Chen. *Feminist Studies*, 39(3): 669–685.

Kieniewicz, J. (2013) Art and science in creative collision [Web blog post], 24th October. Available at: http://blogs.plos.org/attheinterface/2013/10/24/art-and-science-in-creative-collision/ [Accessed 11.2.2015].

Kuhn, L. (2005) A critical reflection on the legitimacy of utilising a complexity approach in social inquiry. Paper presented at the *Complexity, Science and Society Conference*, Liverpool, UK, 11–14th September.

Kuhn, L. (2008) Complexity and educational research: A critical reflection. *Educational Philosophy and Theory*, 40(1): 177–189. doi: 10.1111/j.1469-5812.2007.00398

Latour, B. (2004) Why has critique run out of steam? From matters of fact to matters of concern. *Critical Enquiry*, 30: 225–248. Available at: www.unc.edu/clct/LatourCritique.pdf [Accessed 1.1.2017].

Little Bear, L. (2000) Jagged world views colliding. In M. Batisse (ed.) *Reclaiming Indigenous voice and vision.* Vancouver, BC: University of British Columbia Press. Available at: www.learnalberta.ca/content/aswt/worldviews/documents/jagged_worldviews_colliding.pdf [Accessed 14.10.2019].

Little Bear, L. (2011) *Native science and Western science: Possibilities for collaboration.* Lecture on 4th of March at Arizona State University. Available at: www.youtube.com/watch?v=ycQtQZ9y3lc [Accessed 15.11.2018].

Little Bear, L. (2016) *Blackfoot metaphysics 'waiting in the wings'.* Congress of the Humanities and Social Sciences Big Thinking Lecture, 1st June. Available at: www.youtube.com/watch?v=o_txPA8CiA4 [Accessed 18.11.2018].

Mason, M. (2009) Making educational development and change sustainable: Insights from complexity theory. *International Journal of Educational Development*, 29(2): 117–124. Available at: http://repository.lib.eduhk.hk/jspui/handle/2260.2/4538

Mignolo, W.D. and Tlostanova, M.V. (2006) Theorizing from the borders: Shifting to geo- and body-politics of knowledge. *European Journal of Social Theory*, 9(2): 205–221.

Müller-Merbach, H. (2006) Heraclitus: Philosophy of change, a challenge for knowledge management? *Knowledge Management Research and Practice*, 4(2): 170–171.

O'Connell, E. (2006) *Heraclitus and Derrida: Pre-Socratic deconstruction.* New York: Peter Lang.

Olssen, M. (2008) Foucault as complexity theorist: Overcoming the problems of classical philosophical analysis. *Educational Philosophy and Theory*, 40(1): 96–117. doi: 10.1111/j.1469-5812.2007.00406.x

Osberg, D. (2005) *Curriculum, complexity and representation: Rethinking the epistemology of schooling through complexity theory* (Doctoral thesis).Open University, Milton Keynes.

Osberg, D. (2010) Taking care of the future. *Complexity theory and the politics of education*. Rotterdam: Sense Publishers, pp. 153–166.

Osberg, D. (2015) Learning, complexity and emergent (irreversible) change. In E. Hargreaves (ed.) *Sage handbook of learning*. London: Sage, pp. 23–40.

Osberg, D. and Biesta, G. (2007) Beyond presence: Epistemological and pedagogical implications of 'strong' emergence. *Interchange*, 38(1): 31–51. doi: 10.1007/s10780-007-9014-3.

Osberg, D. and Biesta, G. (2008) The emergent curriculum: Navigating a complex course between unguided learning and planned enculturation. *Journal of Curriculum Studies*, 40(3): 313–328. doi: 10.1080/00220270701610746

Pei, S., Zhang, G. and Huai, H. (2009) Application of traditional knowledge in forest management: Ethnobotanical indicators of sustainable forest use. *Forest Ecology and Management*, 257(10): 2017–2202.

Plato (2007 [380 BCE]) *The republic* (Penguin classic). London: Penguin.

Prigogine, I. (1977) Biographical. *Nobelprize.org. The official site of the Nobel prize*. Available at: www.nobelprize.org/nobel_prizes/chemistry/laureates/1977/prigogi ne-bio.html [Accessed 15.2.2015].

Prigogine, I. (1983) *The rediscovery of time: A discourse originally prepared for the Isthmus Institute, presented to the American Academy of Religion*. www.mountainman.com.au/ilyatime.htm [Accessed 23.2.2014].

Prigogine, I. and Stengers, I. (1984) *Order out of chaos: Man's new dialogue with nature*. London: Heinemann.

Pueyo, S. (2014) Ecological econophysics for degrowth. *Sustainability*, 6: 3431–3483. Available at: www.mdpi.com/2071-1050/6/6/3431/htm [Accessed 19.7.2018].

Robinson, T.M. (1987) *Heraclitus (of Ephesus): Fragments. A text and translation with a commentary*. Toronto: University of Toronto Press.

Sacks, J. (2002) *The dignity of difference*. New York: Continuum.

Semetsky, I. (2008) On the creative logic of education, or: Re-reading Dewey through the lens of complexity science. *Educational Philosophy and Theory*, 40(1): 83–95. doi: 10.1111/j.1469-5812.2007.00409.x

Semetsky, I. and Masny, D. (2013) Introduction: Unfolding Deleuze. In I. Semetsky and D. Masny (eds.) *Deleuze and education*. Edinburgh: Edinburgh University Press, pp. 1–18.

Snaza, N., Sonu, D., Teruman, S.E. and Zaliwska, Z. (eds.) (2016) *Pedagogical matters: New materialisms and curriculum studies*. New York: Peter Lang.

Spivak, G. (1993) *The post-colonial critic: Interviews, strategies, dialogues*. New York: Routledge.

Sterling, S. (2010) Transformative learning and sustainability: Sketching the conceptual ground. *Learning and Teaching in Higher Education*, 5: 17–33. Available at: http://www2.glos.ac.uk/offload/tli/lets/lathe/issue5/Lathe_5_S%0Sterling.pdf

Sundberg, J. (2014) Decolonizing posthumanist geographies. *Cultural Geographies*, 21(1): 33–47.

Todd, Z. (2016) An Indigenous feminist's take on the ontological turn: 'Ontology' is just another word for colonialism. *Journal of Historical Sociology*, 29(1): 4–22.

Vaughan, M. (2007) Introduction: Henri Bergson's 'creative evolution'. *SubStance*, 36(3): 7–24. doi: 10.1353/sub.2007.0051

Watson, T. (2017) Indigenous knowledge systems can help solve the problems of climate change. *@indigenousX*, 7th June. Available at: https://indigenousx.com.au/teila-watson-indigenous-knowledge-systems-can-help-solve-the-problems-of-climate-change/#. w0itztjkjiu [Accessed 13.7.2018]. Also posted in *The Guardian*, 2nd June 2017.

Available at: www.theguardian.com/commentisfree/2017/jun/02/indigenous-knowl edge-systems-can-help-solve-the-problems-of-climate-change [Accessed 13.7.2018].

Wollstonecraft, M. (1989 [1787–1797]) *The works of Mary Wollstonecraft*, 7 vols. Todd, J. and Marilyn Butler, M. (eds.). London: Pickering and Chatto.

Zapatistas Army of National Liberation (2005) *Sixth declaration of the Selva Lacandona*. Lahore: Last Word Books.

3 Recognising and challenging ways of being in the world founded on separation, autonomy and rationality

Introduction

In this chapter I explore and challenge the dominant Western (Eurocentric) worldview in which conceptions of subjectivity are based on *the autonomy of each separate individual to make independent decisions based on reasoning*. I consider the meanings and historical roots of these conceptions and their implications for being in the world together, and, in particular, what this means/ has meant for education. The European Enlightenment is examined at the start of the chapter as it had, and continues to have, a significant influence on Western-style education – an approach to education extending beyond European boundaries through structures and processes such as colonialism/post colonialism. I then explore other ways to approach subjectivity – ways which understand the world, and all within it, as connected/relational and also the ethical implications created by such relationality. I introduce these ideas not as 'blueprints' for subjectivity but rather to show that other ways of being a subject are possible. Once again, I acknowledge that these are not always new ways but ways which have often been 'othered' and denigrated in Eurocentric thinking.

As I began to discuss in Chapter 2, a key development within dominant Western philosophical framings of the world was the (European) Enlightenment. This late seventeenth- and eighteenth-century philosophical movement began with the mechanical framings of the world introduced by thinkers such as Descartes, Newton and Boyle. These approaches emphasised breaking things down into what were understood as unchanging parts which could be moved around and then returned to their original position. Breaking things down into parts enabled a categorisation of these parts – categories to which reasoning could then be applied. The European Enlightenment thus also became known as *The Age of Reason*. The 'truths' which derived from reasoning could then be applied to different situations as 'universal truths'.

As explored in Chapter 2 the European Enlightenment was not homogeneous, experiencing internal and external influences and challenges. However, a key feature of the later European Enlightenment period was that a person was considered to be autonomous, i.e. had a capacity to reason and make their own informed decision without the direction of another. This was a reaction against the notion that 'the people' were incapable of understanding the world for themselves and indeed did not need to understand but just 'obey their betters' (God, the Church, the Sovereign and the aristocracy). There were, however, limits to who was deemed capable of such autonomy and the capacity to reason. Certain categories, for example women and children, were excluded by some (many) thinkers. Autonomy, with its emphasis on self-government, identifies the self as independent from others: a self apart. In addition, in this European 'Age of Reason' thinking was prioritised/valued over action and also over emotion. Descartes' (1637) often-quoted phrase 'I think therefore I am' reflects this.

The European Enlightenment had a wide-ranging influence on political and cultural life from literature to political revolution – an influence which still continues today around the world, including in Western-style education. There were challenges to rational autonomous conceptions of the subject, for example in the thinking/practice of mystics and more recent challenges, examples of which are discussed later in the chapter. However, dominant European Enlightenment conceptions remain pervasive and powerful. The allegorical drawing *Liberty armed with the Sceptre of Reason strikes down ignorance and fanaticism* (see Figure 3.1) illustrates Enlightenment beliefs explored thus far in this chapter.

Figure 3.1 Liberté armée du Sceptre de la Raison foudroye l'Ignorance et le Fanatisme [Liberty armed with the Sceptre of Reason strikes down Ignorance and Fanaticism (author's translation)]

(Engraving) (1793), Chapuy, J. B. (engraver) after Simon Louis Boizot (artist), Collection Michel Hennin.

Source: [Reproduced with permission of Bibliothèque Nationale de France]

Kant, a leading figure in the later European Enlightenment period, defined Enlightenment as the release of the human being 'from his self-incurred tutelage': a tutelage which unnecessarily prevented man from making use of his understanding without the direction of another. His thinking has strongly influenced Western-style education. Kant (1964 [1781]) argued that man's capacity to become a fully rational, autonomous individual is a fundamental part of human nature, part of his essence. Education (the development of one's innate capacity to reason) is needed to bring a person to this state of freedom. Thus, education and *freedom* are intimately connected.

At first view, these seem worthy ideas, yet they also give rise to significant issues. First, whilst the focus on freedom brought ideas of what it is to be a human subject into the Western educational sphere it was a *certain type of subject*. As Biesta (2010) points out, what it is to be a human subject in the Kantian Enlightenment framing, namely, a rational autonomous person, has already been decided before anyone (child) actually appears and has a chance to reveal themselves. Children are to be socialised into this existing rational order.

Second, autonomy and rationality are understood to be the natural capacities of a 'free' person. This introduces an essentialist notion of what it is to be human, moreover one emphasising separation, autonomy and rationality.

Third, within the European Enlightenment framing, children are seen as pre-rational. Rationality is a state to attain as a destination of education. However, this is not the only way to conceive the issue of subjectivity in educational settings. Approached in other ways education has potential to open spaces in and through which unique subjectivities can emerge in the present.

Whilst rational autonomous notions of the subject have been and continue to be a very dominant framing in Western-influenced education, other possibilities for approaching the issue of subjectivity exist. These are now explored although they are not offered as some kind of 'solution' or formula for action. Rather, the aim is to help one to see one's own existing (often unnoticed) worldview; open minds to other possibilities and why these ways are valuable and *encourage the emergence of new ways of being as yet unforeseen and unforeseeable*. I focus on the Indigenous conception of 'all my/our relations', introduced in Chapter 2, and also ideas around relationality within Western (Eurocentric) and other thinking which challenge the dominance of autonomy, separation and reasoning/rationality. Relationality is strongly linked to care and ethics in these various conceptions, and consideration of these connections and implications is included.

All my/our relations

As highlighted in the thinking of Little Bear and De Line 'all my/our relations' is central for many Indigenous peoples. All parts of creation are inextricably linked. This is formally expressed in the KARI-OCA 2 declaration produced by a gathering of approximately 500 Indigenous persons from all over the world at the RIO+20 Earth Summit in 2012. The declaration states that environmental policy must respect:

> the inseparable relationship between humans and earth, inherent to Indigenous Peoples . . . for the sake of our future generations and all of humanity . . . our lands and territories are at the core of our existence – we are the land and the land is us.
>
> (KARI-OCA 2 Declaration 2012: online)

Little Bear (2011, 2016) highlights how Western Enlightenment thinking prioritises 'Progress through Reason', using reason to solve problems through experimentation built on breaking the whole into separate parts and on categorisation. In contrast, in Blackfoot Indigenous thinking and being ecological connection and relationship are key. Little Bear comments that in Blackfoot metaphysics (philosophy), with this ecological relational emphasis, thinking is holistic. It is like 'standing on top of a hill' with a panorama of all the various interconnected elements laid out in front of one. Little Bear also draws attention to the role that language plays in such thinking. For example, English is a language dominated by nouns, which are ideal for categorisation. The language of the Blackfoot people has more emphasis on processes and connections, and this is reflected in, for example, the many different verbs and the ways these can be used/conjugated. He suggests that if people want to understand Blackfoot and other Indigenous metaphysics (philosophies) learning the language of that particular people is an excellent way in.

Through embodied thinking and experiences *in* and *with* Bawaka Country, North East Arnhem Land, Australia, a group of Indigenous women Elders, their female relatives and Indigenous and white women scholars also draw attention to the possibilities opened up by language. These authors propose that the ideas contained within the word *gurrutu* 'illustrate the limits of Western ontologies and possibilities for other ways of being and theorising' (Bawaka Country *et al.* 2016: 456). The authors emphasise that they are not presenting an in-depth ethnographic account of *gurrutu* – a complex kinship system that extends beyond the notions of human kin to kinship with the more-than-human world. This would imply that these 'knowing academics' could 'analyse and wrap up its complex meaning in a definitive, authoritative way' (457). Instead they aim to work in ways which can contribute to dismantling 'a hierarchy of knowledge that would place human-centred, academic understandings as more legitimate than the forever-knowledge of Yolŋu Elders and their more-than-human kin' (457). Through participating in, recollecting and retelling the story of all that is involved in a particular day, a particular process of digging for *ganguri* (yams) framed within *gurrutu*, they invite the readers of their article 'to learn what they can from us about what it might mean to live in a world that is relational, that co-becomes with us and each other, that is knowing even in death' (456). In exploring what happens in the embodied responses between ants who bite, mingling their fluids with a young participant, Nanukala, as she digs for yams, the authors also highlight the ethics that *gurrutu* brings since:

> through this active knowing, this co-becoming, 'an ethics of care and responsibility emerges' (Bawaka Country including Suchet-Pearson *et al.* [2013]). It is not enough to attend to the message, to know the ants, to listen to the thousands' language. *Yolŋu* must also respond and respond with care. . . . And not only humans – it is imperative for everything to attend with care and to take responsibility for their co-becomings. Everything is a part of the web of *gurrutu* and this means everything has an obligation.
>
> (467)

Little Bear (2011, 2016) highlights how Indigenous thinking and being fore-ground what Western thinking terms 'sustainability', although they do not use this term. He comments how many Indigenous ontologies accept that humans operate in a narrow band of ideal conditions within the 'whole picture'. These delicate ecosystem conditions need to be maintained for humans to survive and flourish. Rituals and ceremonies are an important part in renewing this eco-logical foregrounding.[1] Thus, Indigenous thinking is orientated towards 'sus-tainability', whereas Western rationalist thinking is orientated towards growth/expansion. The KARI-OCA 2 Declaration at Rio 20+ expresses this holistic, 'sustainability-orientated' thinking more formally in the following way:

> Our lands and territories are at the core of our existence – we are the land and the land is us; we have a distinct spiritual and material relationship with our lands and territories and they are inextricably linked to our sur-vival and to the preservation and further development of our knowledge systems and cultures, conservation and sustainable use of biodiversity and ecosystem management. . . . Caring and sharing, amongst other values, are crucial to bringing about a more just, equitable and sustainable world.
>
> (KARI-OCA 2 Declaration 2012: online)

Relationship between all in a world where every element is animate enables interaction and communication between elements. For example, for many North American Indigenous people fish have an important knowledge role, as fish have lived on the Earth since before the time of the dinosaurs (Little Bear 2016). Humans can engage with and listen to fish so as to understand better how to live within the delicate ecosystems humans inhabit. This does, however, raise the question in Western minds of *how* one listens to fish. Todd (2015, 2016a, 2016b, 2017, 2018), a Métis scholar, biologist and fish philoso-pher brings Western academy thinking into contact with the knowledge of fish. She discusses how initially:

> even with fish woven so intimately into every part of my life, it had never occurred to me that fish were also citizens, interlocutors,

storytellers, and beings to whom I owed reciprocal legal-governance and social duties. These lessons had been deeply erased from dominant (non-Indigenous) public discourse in Alberta and I had not recognised the implicit ways fish were woven into my own life as more than food. This is the thing about colonisation: it tries to erase the relationships and reciprocal duties we share across boundaries, across stories, across species, across space, and it inserts new logics, new principles, and new ideologies in their place.

(2016b: online)

Todd (2016b: online) reflects on her experiences when carrying out research in Paulatuuq. An elder, Annie Illasiak, repeated to her on numerous occasions the teaching that 'you never go hungry in the land if you have fish'. Initially she thought that this teaching was purely a 'utilitarian subsistence or survival lesson: even if every other source of food is unavailable, if you have fish, you won't starve'. However, it was only when she was writing up her research that she began to 'untangle' her utilitarian understandings and started to see the 'fish pluralities' Annie was speaking about. Todd came to understand that:

This was not a lesson solely about food, but about the many manifestations and articulations of human-fish relations in Paulatuuq: as long as you have fish, you have stories, memories and teachings about how to relate thoughtfully with the world and its constituents. As long as you have fish (and other animals), you are nourished not only physically, but in a plurality of emotional, spiritual and intellectual ways as well. A world without fish is not only a hungry one, but one intellectually and socially bereft.

(online)

It is through 'tending' within 'communities of care' to Indigenous laws,[2] stories, memories and teachings and engaging in the various practices (which are reinforced/renewed through ceremonies and stories) that the people of *Paulatuuq can attend to, and care for fish.* Todd reflects on the need to bring such caring to academic work in the Western Academy. She comments that she has 'been thinking a lot lately about what it means to inhabit space and time with care and tenderness', asking what it 'means to tend to and mobilise communities of care, to embody ethics of care, reciprocity and kindness in our work' (2018: 74). To explore these issues she draws on Donald's (2009) thinking on ethical relationality (also see Donald *et al.* 2012). Donald *et al.* highlight the importance of being in 'relation' with the community in which we live, work and research; the ethical 'responsibilities that come with being in relation' (71) and the necessary responses to these which so easily become absent from one's private and public life.

'Listening to fish' and approaching all in the world as having agency does raise possible accusations of anthropocentrism, anthropomorphism and theriomorphism (ascribing animal characteristics to humans). De Line (2016) responds to this criticism by highlighting how 'stating that all matter is made up of spirit and that spirit is what is scientifically called energy waves in all my/our relations is not anthropocentric, anthropomorphic nor theriomorphic since it 'places no grid or hierarchy among what is defined as *all*' (6).

Engaging with Indigenous Science, ideas of flux and 'all my/our relations' can be challenging for Western minds. Education has potential to be a place to explore these other ways of knowing and being in the world and the challenges presented. An important point to make here is that rather than focusing on which ideas are 'true' or 'correct', education can value the possibilities that bringing these different ways of knowing, being and acting in the world into conversation with each other may open up. Education exploring these ideas also needs to acknowledge the harms that Western suppression of them has caused/continues to cause. In later chapters, I introduce a variety of approaches to encourage such educational processes.

Education can also be a place to acknowledge and explore that whilst dominant Western (Eurocentric) thinking foregrounds universal Reason, autonomy, separation and categorisation, other ways of thinking have also developed within 'Western' and other ontologies. It is therefore to some of these ways that this chapter now turns.

Relationality in Western and other ontologies

The thinking of Buber, Macmurray and a range of feminist thinkers all foreground relationality.[3] For these various thinkers, relationality, as in Indigenous ontologies/framings of the world, is profoundly interconnected with care and ethical considerations.

Martin Buber's I-Thou relationships and John Macmurray's action and relationality

Whilst popular, and often explored within Western thinking, Buber was a Jewish philosopher and educator. Buber's own education encompassed studies of both Jewish/Hebraic and Western (Hellenic/Greek) philosophy. Writing from the 'borderlands' of the two traditions, he developed ideas which critique and go beyond Eurocentric worldviews built on Greek (Hellenic/Western) traditions. In *I and Thou* (1958 [1923]), a series of thoughts or aphorisms, he introduces what he calls the I-It relationship. In this I-It framing, modern (i.e. European Enlightenment) man is supposedly an 'objective observer' (rather than an active participant within the world). This 'observer' collects data and analyses, classifies and theorises about 'It' (the Western approach also noted by Little Bear). The object of observation – the 'It' – is approached as a thing to be known and utilised. Buber then proposes a different way to encounter the other, in an *attentive* 'I-Thou' relationship. The 'I' engages with 'Thou' in its entirety not just as a sum of qualities. The encounter fills the universe and is not merely a point in space or time. It is transformative of both. As Buber (1958: 24–25) proposes:

> The primary word *I-Thou* can be spoken only with the whole being. Concentration and fusion into the whole being can never take place through my agency, nor can it ever take place without me. I become through my relation to the *Thou*; and as I become the *I*, I say *Thou*.
>   All real living is meeting (encounter).

Smith (2009: online) points out that for Buber 'Encounter' (*Begegnung*) is an event or situation in which relation (*Beziehung*) occurs. Smith continues, 'We can only grow and develop, according to Buber, once we have learned to live in relation to others, to recognise the *space* between us, and build a society based on *I-Thou* relationships. The fundamental means is dialogue'. Such an emphasis on a *space* between the two in dialogue, and the potential that this can open up particularly in relation to education, is fundamental to the ideas I develop in this book. *Such space allows the other to exceed our (preconceived) objectifying idea of them*. As Levinas (2000 [1961]: 51) puts it:

> To approach the other in conversation is to welcome his expression, in which at each instant he overflows an idea a thought would carry away with it. It is therefore to receive from the other beyond the capacity of the I, which means exactly: to have the idea of infinity.

Buber does not limit *I-Thou* encounters to encounters between humans. For Buber *I-Thou* encounters with the wider natural world are also possible. He reflects that in Western thinking there are a variety of ways of understanding a tree as an object (e.g. as a resource), as an object of contemplation or as an example of a particular species category:

> But it can also happen, if will and grace are joined, that as I contemplate the tree I am drawn into a relation, and the tree ceases to be an it – a mere example of a tree, but instead it is with this tree, on this occasion that I engage into a relation and this fills my world.
>
> (59)

To address the question of whether a tree has consciousness similar to our own, Buber comments 'I have no experience of that. But thinking that you have brought this off in your own case, must you again divide the indivisible? What I encounter is neither the soul of a tree nor a dryad, but the tree itself' (59).

  Buber's I–Thou thinking has been influential in education (for example see discussion in Smith 2009), opening ways to enter into attentive, transformative encounters with specific others. Buber's ideas also offer a way for individual teachers to resist, at a personal level, the 'commodification' of education and the objectification of individuals (i.e. where individuals are approached as an 'it' rather than a 'thou') dominant in much education today.

The Scottish philosopher John Macmurray also challenges Western framings of the world and of subjectivity within such a framing, foregrounding instead action and relationality. He contests Kant's' solitary (autonomous) ego and the prioritisation of mind (thinking) over body – as in Descartes' 'I think therefore I am' and critiques the dualism (i.e. the separation of mind and body) that this prioritisation creates.

Macmurray (1957, 1991 [1960], 2004) proposes action as a starting point (as primary) since action involves both the mind and the body. For Macmurray, 'what we do', our actions, inaugurates the future. In addition, thinking is itself an action, since whilst we are 'doing' we are also reflecting on our doing. Such reflection on action can then lead to further action informed (and made better) by reflection. He acknowledges, however, that sometimes problems emerge which make it difficult or impossible for us to go on as we are. Philosophy's role can then be to withdraw from action for a period of time, responding to the need to stop and think: opening space and time to 'imagine different ways of addressing our problem, and to consider the consequences of choosing one of them' (Godway 2010: 1). Like Macmurray, the political thinker Hannah Arendt (whose concepts are central to the discussions in later chapters) challenges the placing of contemplation above action in Western metaphysics. She also recognises that thinking is itself an action (see Arendt's *Life of the Mind*, 1977 [1971]). Both thinkers thus contribute ways to challenge dominant Western philosophical framings, opening up possibilities for new ways of being and acting.

Macmurray further develops his ideas to argue that human motivation is towards what he calls the *personal*: a desire for relationship. He contrasts this with two other modes – the mechanical and the organic: modes he feels have come to dominate Western (Eurocentric) thinking. For Macmurray, 'the mechanical' is thinking informed by 'supposedly-objective' experimental physics operating within an unfolding universe. Organic thinking does recognise embodiment but lacks recognition of the spiritual and emotional aspects of being (see Costello 2002 for further discussion). Macmurray looks to the early relationship between a mother and child for evidence of the desire towards a relationship. His ideas on the personal are brought together in his comments: 'All meaningful knowledge is for the sake of action, and all meaningful action for the sake of friendship' (Macmurray 1991: 15).

In educational settings, Macmurray's ideas can help focus attention on the importance and value of action, embodiment and relationship. In addition, space can be opened to explore when action needs to make way for reflection on issues which seem intractable to current solutions. This can create possibilities for allowing emergence of new futures not 'colonised' by ideas from the present (Facer 2016) but instead left radically open.

Another important resistance to dominant Western ontologies has been developed by feminist writers interested in relational rather than autonomous rational ways of being together and caring for each other in the world. These ideas are the focus of the next section of this chapter.

Feminist relational ethics of care

Carol Gilligan is a landmark thinker in the development of feminist ethics of care. In the 1980s, she critiqued the American psychologist Kohlberg's (1958) model of what he identified as stages of moral development. In his model the highest level of moral development is based on acquiring and using rationality and abstract 'universal principles' to make ethical decisions (as in Kantian approaches to ethics). According to Kohlberg, males demonstrated a stronger tendency to move through the hierarchy to a position of using rationality and universal principles when faced with ethical decisions. Gilligan was dissatisfied with Kohlberg's model since she observed that females tended to focus on responding to particular situations. As this is only at the middle stages of Kohlberg's model it placed women and girls in a deficit position compared with similarly aged males. Gilligan (1982: 484) highlighted how:

> The very traits that have traditionally defined the goodness of women, their care for and sensitivity to the needs of others, are those that mark them out as deficient in moral development.

Gilligan carried out her own research, publishing it as *In a Different Voice* (1982). This is considered by many as a seminal work, marking a shift to 'difference feminism' which 'represents a broad spectrum of feminisms that emphasise differences between women and men. This approach arose in the 1980s and 1990s in efforts to revalue qualities traditionally devalued as "feminine" such as subjectivity, caring, feeling or empathy' (Gendered Innovations: online). In this she argues that women are not deficient in moral reasoning but instead use a style of reasoning not valued by Kohlberg. Women and girls have a stronger tendency towards an 'ethics of care' in which ethical decisions respond to, and build from, caring for others in specific situations rather than drawing on 'seemingly' universal reason and codes of behaviour. Whilst this 'ethic of care' is not itself limited to females, she argued it was more common amongst her female participants. 'Difference feminism' was controversial. For example, some thinkers accused it of essentialism i.e. framing women as having certain 'essential' characteristics and tendencies (see Ball 2010 for further discussion).

Engaging with the work of Gilligan, but working with a more phenomenological approach (i.e. how care is actually experienced), Nel Noddings (2013) also identified relationships and the relational (i.e. concerning/arising from interconnection) rather than universal principles, as the starting point for ethics.[4] For Noddings relationality is a fundamental part of being human (one's ontology) and recognising human encounter and affective response is a key aspect of human existence.

Like Gilligan, Noddings rejects notions of universal ethics that draw on supposedly rational objectivity, what she calls 'a thinking mode that moves the self toward the object. It swarms over the object, assimilates it' (33). Instead Noddings emphasises the uniqueness of each human in the encounter between the *one-caring* and the *cared-for*. In caring relations, the one-caring is open, '*receptive to the other*', '*engrossed*' in the other, and it is through these encounters that ethical responses emerge. Such caring relationships are characterised by 'warm acceptance and trust' (65) by the one-caring of the '*cared-for* as he is' (65). This acceptance and trust are what Noddings calls *confirmation*. For Noddings 'Caring is *completed* in all relationships through the *apprehension* (recognising) of the caring by the cared-for' (65) which generates a form of *reciprocity*.

Noddings uses the term 'inclusion' to describe the situation in which the one-caring tries to see with two pairs of eyes – their own and those of the one cared-for. This is not to say that thinking/analysing has no part in the process, but it is not the starting point. She comments:

> In the language of Martin Buber, the cared-for is encountered as "Thou", a subject, and not as 'It', an object of analysis. During the encounter, which may be singular and brief or recurrent and prolonged, the cared-for 'is "Thou" and fills the firmament.
>
> (176)

Noddings provides an example from teaching mathematics. If one encounters a student who does not like mathematics, or even fears it, then rather than work from one's own position as a lover of mathematics and try to communicate that to the student through interesting activities, different explanations, etc., the teacher puts herself instead in the position of the student. She asks herself what it must be like to fear mathematics or find it uninteresting, what might cause that and how could those feelings be responded to and potentially overcome. This contributes to the development of caring and productive relationships (as well as contributing to improvements in the students' knowledge of mathematics). Teachers should have opportunities to work over extended periods of time with students to build up confirmation – warm acceptance and trust. Curriculum subjects can and should have importance in education but not at the cost of the development of caring relationships which can then inform caring relationships in the world at large.

Critics of Noddings question her focus on a particular type of mother/child relationship, since mothering and child-rearing is different in different cultures (as well as within cultures) (Hassan 2008). Held (2007), however defends taking 'care' as a starting point for ethics, saying that we have all received care of some kind, as otherwise we would not have lived through babyhood/childhood. Thus, care is universally *experienced* even if the particular experiences and providers of care may differ. Other critics (for example see discussion in Hassan 2008) argue that Noddings' approach encourages traditional gender roles for women, ignores virtues besides care and is too unidirectional (Hoagland 1990). They also argue that Nodding's ethics of care could be improved by adding autonomy and justice into the theory. In exploring this point, Held (1995a, 1995b, 2007) argues that relationality and care should not be held in juxtaposition with universal notions of justice which emphasise universal 'rights' such as equality, civil rights and self-determination (autonomy). Rather, care can be understood as the wider moral framework into which justice should be fitted. For Held, 'Care seems to me the most basic moral value. . . . Within a network of caring, we can and should demand justice, but justice should not then push care to the margins' (1995a: 131). Held also highlights that care should not be limited to the private domain of family. Held is able to hold relationality and autonomy together by asserting, 'Persons are relational and interdependent. We can and should value autonomy, but it must be developed and sustained within a framework of relations of trust' (1995a: 132). She argues that ethics of care 'conceptualise persons as deeply affected by, and involved in, relations with others' (2007: 44) and recognises how for many care theorists 'persons are at least partly constituted by their social ties' (44).

Another important criticism of Noddings' ethics of care is that it limits caring relations to those between humans. In her theory, for relations to be caring they must somehow be apprehended and thus completed by the other. Noddings acknowledges that encountering the other-than-human can be beneficial to the wellbeing of the one-caring. However, for Noddings, the other-than-human cannot apprehend the caring and respond in a *reciprocal* caring relationship. Noddings acknowledges her Western scientific training means that for her, whilst talking to her plants and tending to them might sustain her, the flowers do not have agency to respond to her with reciprocal caring, and to think otherwise is anthropomorphic. However, as explored earlier in the chapter, for many Indigenous people the intra-connection between human and other-than-humans is central, and agency is not limited to the human realm. For example, the Anishinaabe scholar and activist Deb McGregor (2009: 37) highlights how in Anishinaabe thinking water is not understood as inert. Instead water has a caring role and:

> We must look at the life water supports (e.g. plants/medicines, animals, people, birds *etc.*) and the life that supports water (e.g. the earth, the rain, the fish). Water has a role and a responsibility to fulfil just as people do. We do not have the right to interfere with water's duties to the rest of creation. Indigenous knowledge tells us that water is the blood of Mother Earth and that water itself is considered a living entity with just as much right to live as we have.

Different understandings and ways of relating to, and caring for, the wider natural world have also been developed within Western feminist thinking. For example, Tronto and Fisher (1991: 40) propose a relational rather than universal approach to caring which does include caring relations between humans and other-than-humans, commenting:

> On the most general level, we suggest caring be viewed as a species activity that includes everything that we do to maintain, continue and repair our 'world' so that we can live in it as well as possible. That world includes our bodies, ourselves, and our environment, all of which we seek to interweave in a complex, life-sustaining web.

For Tronto and Fisher, such caring is not dyadic (interaction between two people, e.g. mother and child) or individualistic and finds a number of ways of expression. It does not 'romanticise' the mother-child relationship, particularly the Western presentation. Care can be single acts or a process, part of the 'web of life'. Tronto and Fisher's definition opens a way to explore the possibility of relationality and care beyond the limitation of the human realm.

Puig de la Bellacasa (2017) draws on Tronto and Fisher's definition to explore the possibility of relationality and care in ways which extend beyond using words and Western rationality and which recognise the 'liveliness of things'. She emphasises that her project is *speculative* rather than normative, i.e. she is interested in provocations and exploring possibilities rather than arguing for particular ways care should be and how it should be practiced. Puig de la Bellacasa speculatively proposes 'touch' (the haptic) as a way to shift from abstraction and detachment to engage with the supposedly mundane and everyday. As Ticktin and Wijsman (2017: online) highlight, for Puig de la Bellacasa, touch 'is a way to transform our thick present and the futures we want to co-create. It is not aimed at more accurately knowing a 'real' world, but at more involvement and commitment to it'. There is resonance between the ideas of Tronto and Fisher and Puig de la Bellacasa and those of Bawaka Country *et al.* (2016), where the authors highlight that in finding and harvesting *ganguri* (yams) '*attending*' to the messages is important. The messages tell us:

> what to do and make us who we are. The women hear the *gukguk* and they remember to take water; they shoo away the flies and know it is *Midawarr*. A certain configuration of knowing, doing, being, telling emerges. Everything is entangled in a web of connectivity which is constantly in motion, constantly co-becoming.
>
> (462)

The feminist theorist Donna J. Haraway also foregrounds entanglement in her thinking on relations to the earth and all its inhabitants. Rather than adopting the term 'Anthropocene' to describe our current era of ecological crisis, Haraway has created the term 'the Chthulucene'[5] (2015, 2016). For her, 'it more aptly and fully describes our epoch as one in which the human and non-human are inextricably linked in tentacular practices' (2016: abstract) in the 'thick present' (1). For Haraway, a central feature of the Chthulucene is that it requires *sym-poiesis*, or making-with, rather than *auto-poiesis*, or self-making. Sym-poiesis demands a 'staying with the trouble', since it is '[l]earning to stay with the trouble of living and dying together on a damaged earth' that 'will prove more conducive to the kind of thinking that would provide the means to building more liveable futures' (abstract). This line of thinking opens a role for education as a place to 'stay with the trouble' and to explore how such 'staying with' can open up new futures. Rosiek *et al.* (2020) highlight the importance of engaging with incommensurabilities as well as commensurabilities in differing worldviews when 'staying with the trouble' but argue that respectful engagement with these is both important and possible. They point out that not engaging in such processes through fear of misunderstanding or accusations of appropriation reinforces and continues the silencing of other ways of thinking and being which are currently excluded from structures and places, including educational ones, dominated by particular Western philosophical conceptions.

In reflecting on the future, Adam and Groves (2011) focus on expanding the reach of relationality and care to consider not only our entanglement with humans and the wider natural world in the present but also inter-/intra-connection and dependence with(in) the future. They propose that 'part of our sense that our lives are going well now is bound up with our expectations for the ongoing futures of what we care about, even beyond our own deaths' (24). This is particularly important in an era in which humans are now making permanent and harmful impacts on the planet we share with others (Jonas 1984). Rather than contractual reciprocity between autonomous individuals, which must necessarily take place in the present, Adam and Groves (2011) argue for what they call 'future-tended care'. Such care reaches out from 'tending' in the present in individual, social, institutional and ecological settings to 'reaching toward posterity, toward futures which already exist *in potentia* within the living futures that are underway within our actions in the present' (25). 'Tending' asks for positive desire and action to 'tend to the well-being of what matters to us' and to 'extending ethical activities' (25) rather than limiting activities based on a fear of consequences.

This chapter has introduced and challenged dominant Western framings of the rational autonomous subject. I have explored other ways of understanding subjectivity – ones which draw on relationality and entanglement, both with other humans and with the wider web of life in the world we share. I have drawn these ideas from a range of sources and acknowledge that many of these have existed for millennia. I share these ideas not as 'blueprints' for the creation of new supposed 'universal' ontologies and epistemologies. Rather, I have shared these ideas to highlight that other ways of understanding and being a subject in the world *are* possible and also to learn with and from them. The next step in the journey shared through this book is to look at ways to encourage, especially in educational settings, the opening of spaces where emergence of such unique subjects is possible.

Notes

1   *The Anchorage Declaration* (Indigenous People's Global Summit on Climate Change 2009), written by Indigenous representatives from the Arctic, Russia, the Americas, the Caribbean, the Pacific, Asia and Africa, also reflects this.
2   For further discussion of concept of Indigenous law and why use of this term is important, see Donald (2009), Todd (2016b, 2016c, 2018).
3   I use the word 'relationality' to express an approach which emphasises how one is characterised/constituted in and through one's relationships with others rather than an approach emphasising autonomy and separation.
4   Noddings first published *Caring: A Feminine Approach to Ethics and a Moral Education* in 1984. This seminal work was republished in 2003 and then in 2013, with the new title *Caring: A Relational Approach to Ethics and a Moral Education*. In the preface to the 2013 edition, she explains how she originally chose 'feminine' as opposed to 'women's' approach, as 'feminine' can include men engaging in traits or behaviours traditionally associated with women. However, she recognises that some find the notion of 'feminine' off-putting and/or problematic since it could appear to suggest that caring is an essential part of women's subjectivity – a positioning she strongly refutes in her work. She argues that 'feminine' is a mode of experience . . . and men might also share this experience' (2013: xxiv). Thus, 'relational' seems a better word to choose, since virtually all care theorists make 'the relation' more fundamental than the individual (xiii)'.
    Swanson (2015: 99), who identifies as a Métis healthcare leader, supports Noddings' defence against accusations of essentialism. Swanson argues that 'going beyond imposed constructs of "feminine" to contemporary understanding of feminine as characteristics that are not the sole domain of women' (99) moves Noddings' thinking beyond accusations of essentialism 'to a place in which our relatedness to each other and Mother Earth' (99) can be recognised and celebrated. Swanson highlights that the ecofeminist ethic of caring which she identifies in Noddings' writing 're-attaches our collective consciousness to the timeless wisdom of Indigenous and First Nations peoples, who knew themselves to be inextricably linked to Mother Earth' (100) – a positioning discussed earlier in this chapter and in Chapter 2.
5   Haraway (2016: 1) explains how her term 'Chthulucene' is:

> a compound of two Greek roots (*khthôn* and *kainos*) that together name a kind of timeplace for learn-ing to stay with the trouble of living and dying in response-ability on damaged earth. *Kainos* means now, a time of beginnings, a time for on-going, for freshness. Nothing in *kainos* must mean conventional pasts, presents, or futures. There is nothing in times of beginnings that insists on wiping out what has come before, or, indeed, wiping out what comes after. *Kainos* can be full of inheritances, of remembering, and full of comings, of nurturing what might still be. I hear *kainos* in the sense of thick, ongoing presence, with *hyphae* infusing all sorts of temporalities and materialities.
>   Chthonic ones are beings of the earth, both ancient and up-to-the- minute. I imagine chthonic ones as replete with tentacles, feelers, dig-its, cords, whiptails, spider legs, and very unruly hair. Chthonic ones romp in multi-critter humus but have no truck with sky-gazing *Homo*. Chthonic ones are monsters in the best sense; they demonstrate and perform the material meaningfulness of earth processes and critters.

References

Adam, B. and Groves, C. (2011) Futures tended: Care and future-oriented responsibility. *Bulletin of Science, Technology & Society*, 31(1): 17–27.

Andreotti, V. (2016) The educational challenge of imagining the world differently. *Canadian Journal of Development Studies/Revue canadienne d'études du développement*, 37(1): 101–112.

Arendt, H. (1977 [1971]) *The life of the mind*. New York: Harcourt, Brace, Jovanovich.

Ball, L. (2010) Profile of Carol Gillingham. In A. Rutherford (ed.) *Psychology's feminist voices multimedia internet archive*. Available at: www.feministvoices.com/carol-gilligan/ [Accessed 19.11.2018].

Bawaka Country, Wright, S., Suchet-Pearson, S., Lloyd, K., Burarrwanga, L., Ganambarr, R., Ganambarr-Stubbs, M., Ganambarr, B., Maymuru, D. and Sweeney, J. (2016) Co-becoming Bawaka: Towards a relational understanding of place/space. *Progress in Human Geography*, 40(4): 455–475.

Biesta, G. (2010) *Good education in an age of measurement: Ethics, politics, democracy*. London: Paradigm Publishers.

Buber, M. (1958 [1923]) *I and thou*. Kaufmann, W. (tr.). New York: Scribner.

Costello, J. (2002) *John Macmurray: A biography*. Edinburgh: Floris Books.

De Line, S. (2016) All my/our relations: Can posthumanism be decolonized? *Open! Platform for Art, Culture & the Public Domain*. Available at: www.onlineopen.org/all-my-our-relations [Accessed 23.1.2018].

Descartes, R. (1960 [1637]) *Discourse on method*. Lafleur, L.J. (tr.). New York: The Liberal Arts Press.

Donald, D. (2009) Forts, curriculum, and Indigenous Metissage: Imagining decolonisation of Aboriginal-Canadian relations in educational contexts. *First Nations Perspectives*, 2(1): 1–24.

Donald, D., Glanfield, F. and Sterenberg, G. (2012) Living ethically within conflicts of colonial authority and relationality. *Journal of the Canadian Association for Curriculum Studies*, 10(1).

Facer, K. (2016) Using the future in education: Creating space for openness, hope and novelty. In H.E. Lees and N. Noddings (eds.) *The Palgrave international handbook of alternative education*. London: Palgrave Macmillan, pp. 53–78.

Gendered Innovations (online) *Feminisms*. Available at: https://genderedinnovations.stanford.edu/terms/feminism.html#:~:text=Difference%20Feminism%20represents%20a%20broad,caring%2C%20feeling%2C%20or%20empathy [Accessed 5.7.2020].

Gilligan, C. (1982) *In a different voice: Psychological theory and women's development*. Cambridge, MA: Harvard University Press.

Godway, E.M. (2010) The crisis of the personal: Macmurray, postmodernism, and the challenge of philosophy today. *Appraisal*, 8(1): 1–15. Available at: http://johnmacmurray.org/wp-content/uploads/2012/02/godway.pdf [Accessed 11.12.2018].

Haraway, D. (2015) Anthropocene, Capitalocene, Plantationocene, Chthulucene: Making kin. *Environmental Humanities*, 6: 159–165.

Haraway, D. (2016) *Staying with the trouble: Making kin in the Chthulucene*. Durham, NC: Duke University Press.

Hassan, T. (2008) *An ethic of care critique* (Digital repository). SUNY Press, pp. 159–162. Available at: http://dspace.sunyconnect.suny.edu/handle/1951/43954 [Accessed 21.11.2018].

Held, V. (1995a) The meshing of care and justice. *Hypatia*, 10(2): 128–132.

Held, V. (1995b) *Justice and care: Essential reading in Feminist Ethics*. Boulder, CO: Westview Press.

Held, V. (2007) *The ethics of care: Personal, political, global*. Oxford: Oxford University Press.

Hoagland, S.L. (1990) Some concerns about Nel Noddings' "Caring". *Hypatia*, 5(1): 109–114.

Indigenous People's Global Summit on Climate Change (2009) *The Anchorage Declaration*. Available at: https://unfccc.int/resource/docs/2009/smsn/ngo/168.pdf [Accessed 10.3.2020].

Jonas, H. (1984) *The imperative of responsibility: In search of an ethic for the technological age*. Chicago: The University of Chicago Press.

Kant, I. (1964 [1781]) *The critique of pure reason*. Kemp Smith, N. (tr.). London: Palgrave Macmillan.

KARI-OCA 2 Declaration (2012) *KARI OCA 2 declaration: Indigenous peoples global conference on RIO + 20 and mother earth*. Accepted by Acclamation, Kari-Oka Village, at Sacred Kari-Oka Púku, Rio de Janeiro, Brazil, 17th June. Available at: https://wrm.org.uy/other-relevant-information/kari-oca-2-declaration-indigenous-peoples-global-confer ence-on-rio-20-and-mother-earth/ [Accessed 17.12.2019].

Kohlberg, L. (1958) *The development of modes of moral thinking and choice in the years ten to sixteen* (Unpublished doctoral dissertation). University of Chicago, Chicago.

Levinas, E. (2000 [1961]) *Totality and infinity. An essay on exteriority*. Lingis, A. (tr.). Pittsburgh, PA: Duquesne University Press.

Little Bear, L. (2011) *Native science and Western science: Possibilities for collaboration*. Lecture on 4th of March at Arizona State University. Available at: www.youtube.com/watch?v=ycQtQZ9y3lc [Accessed 15.11.2018].

Little Bear, L. (2016) *Blackfoot metaphysics 'waiting in the wings'*. Congress of the Humanities and Social Sciences Big Thinking Lecture, 6. Available at: www.youtube.com/watch?v=o_txPA8CiA4 [Accessed 18.11.2018].

Macmurray, J. (1957) *The self as agent*. London: Faber and Faber.

Macmurray, J. (1991 [1960]) *Person in relations*. London: Humanities Press International.

Macmurray, J. (2004) *Selected philosophical writings*. McIntosh, E. (ed.). Exeter: Imprint Academic.

McGregor, D. (2009) Honouring our relations: An Anishnaabe perspective on environmental justice. In J. Agyeman, P. Cole and R. Haluza-Delay (eds.) *Speaking for ourselves: Environmental justice in Canada*. Vancouver, BC: University of British Columbia Press, pp. 27–41.

Noddings, N. (2013) *Caring: A relational approach to ethics and moral education*. Oakland: University of California Press.

Puig de la Bellacasa, M. (2017) *Matters of care: Speculative ethics in more than human worlds*, 3rd Edition. Minneapolis: University of Minnesota Press.

Rosiek, J., Snyder, J. and Pratt, S. (2020) The new materialisms and Indigenous theories of non-human agency: Making the case for respectful anti-colonial engagement. *Qualitative Inquiry*, 26(3–4): 331–346.

Smith, M.K. (2009) Martin Buber on education. *The Encyclopedia of Pedagogy and Informal Education*. Available at: https://infed.org/mobi/martin-buber-on-education/ [Accessed 3.4.2020].

Suchet-Pearson, S., Wright, S., Lloyd, K., Burarrwanga, L. and Hodge, P. (2013) Footprints across the beach: Beyond researcher-centred methodologies. In J.T. Johnson and S.C. Larsen (eds.) *A deeper sense of place: Stories and journeys of collaboration in indigenous research*. Corvallis: Oregon State University Press, pp. 21–40.

Swanson, L. (2015) A feminist ethic that binds us to Mother Earth. *Ethics and the Environment*, 20(2): 83–103.

Ticktin, M. and Wijsman, K. (2017) *Review: Maria Puig de la Bellacasa – Matters of care: Speculative ethics in more than human worlds*. Available at: https://www.hypatiareviews.org/reviews/content/337 [Accessed 21.1.2020].

Todd, Z. (2015) Decolonial dreams: Unsettling the academy through "*namewak*". A contribution on appropriation of Indigenous voice and human-fish relations. In C. Picard (ed.) *The new (new) corpse*. Chicago, IL: Green Lantern Press.

Todd, Z. (2016a) An Indigenous feminist's take on the ontological turn: 'Ontology' is just another word for colonialism. *Journal of Historical Sociology*, 29(1): 4–22.

Todd, Z. (2016b) From fish lives to fish law: Learning to see Indigenous legal orders in Canada. *Somatosphere*, 1st February. Available at: http://somatosphere.net/2016/02/from-fish-lives-to-fish-law-learning-to-see-indigenous-legal-orders-in-canada.html [Accessed 6.12.2018].

Todd, Z. (2016c) From a fishy place: Examining Canadian state law applied in the Daniels decision from the perspective of Métis legal orders (invited piece). *TOPIA*, 36(Fall): 43–57.

Todd, Z. (2017) Fish, kin, and hope: Tending to water violations in *amiskwaciwâskahikan* and treaty six territory (invited piece). *After all: A Journal of Art, Context and Inquiry*, 43(1): 102–107.

Todd, Z. (2018) Refracting colonialism in Canada: Fish tales, text, and insistent public grief (invited piece). In M. Jackson (ed.) *Coloniality, ontology, and the question of the posthuman*. London: Routledge, pp. 131–146.

Tronto, J.C. and Fisher, B. (1991) Toward a feminist theory of caring. In E. Abel and M. Nelson (eds.) *Circles of care*. Albany, NY: SUNY Press, pp. 36–54.

4 Opening the possibility of emergence of other ways to be a human subject

A role for 'spaces of appearance'

Introduction

In this chapter I focus in more depth on the intricacy of thinking about human subjectivity, and the possibility of opening spaces in which other *as yet unforeseen and unforeseeable subjectivities can emerge in educational settings*: subjectivities with potential to challenge dominant Western (Eurocentric) philosophies which frame the world as stable and separate and which decide in advance that the subject is a rational autonomous being. I welcome how ways of being a subject can be *informed* by the existing relational ideas of subjectivity explored in Chapter 3. However, here I focus on the possibility of opening spaces in and through which *new subjectivities as yet undetermined* can appear.

Before looking at the possibility of opening such 'spaces of appearance' (Arendt 1974 [1958]: 199), it is first helpful to examine the terms 'subject' and 'subjectivity' in dominant Western thinking in more detail. Second, it is helpful to examine some of the problems that the very act of attempting to think about subjectivity creates. I include, as ways to address the problems generated by this act of *theorising subjectivity*, the ideas of two very different contemporary Western thinkers who have addressed this – those of Biesta (2006, 2010, 2013) in his 'pedagogy of interruption' and Braidotti's (2011a, 2013b, 2016) concept of the 'nomadic subject'. I then identify aspects from these thinkers which I can take forward into my own engagement with opening 'spaces of appearance' in and through which new subjectivities can emerge. I also reflect how engagement can be enriched by the Indigenous and other ontologies introduced in Chapters 2 and 3. This is highly pertinent to education, for, as Arendt (2006a [1961]: 193) points out, education is where we need to encourage possibilities of *not* 'striking from the hands of children their chances of undertaking something new, something unforeseen by us'.

Subjectivity

As introduced in Chapter 1, the Western philosophical terms 'subject' and 'subjectivity' have a sense of having consciousness of the self as a being with agency and with capacity to make choices in the world. The terms 'subject' and 'subjectivity' also imply awareness of the self as having the ability to have higher-order thoughts and reflect on one's own 'beingness'. In addition, as Biesta (2010, 2013) reminds us, in Western framings the word 'subject' also has the sense of one who acts and one who is acted upon (subjected to).

In dominant Western thinking, subjectivity, the capacity for consciousness, is reserved for humans, reinforcing an *anthropocentric* view of subjectivity which places humans as superior to others in the world. The issue of whether animals, and other parts/participants in the wider natural world have this dominant Western sense of consciousness (the capacity to have higher order thoughts and reflect on the self) is a current topic of debate in academic literature. In relation to the possibility of animal subjectivity, some writers argue, at one end of the scale, that animals have no rationality and no cognitive capacities but instead adapt, via the mechanisms of natural selection, over a number of generations. For example, Carruthers (1998) proposes that higher order thoughts and consciousness are not a feature of animals since animals cannot, for example, contemplate a colour in the abstract but only experience it physically. At the other end of the scale, Lyvers (1999) argues for the possibility of animal subjectivity. To support his argument Lyvers (1999: 5) states:

> The attribution of subjective states to non-human animals is sometimes derided as "anthropomorphism" but simple acknowledgment of animal subjectivity is not equivalent to assuming that an animal's conscious experience must be similar in all respects to one's own.

Recent thinking about plants and minerals also challenges notions that these have no possibility of subjectivity and adapt purely through evolution, albeit extending Lyver's point, their subjectivity does not have to be the same as/limited to current Western definitions of subjectivity. The biologist Jean-Marie Pelt (2004) explores the solidarity and community which exists within the wider natural world. For example, he challenges the assumed boundaries between the categories of animal, vegetable and mineral, citing as an example coral reef ecosystems where these boundaries are blurred. In *The Hidden Life of Trees* (2016) the forester Peter Wohlleben explores the community, communication and support mechanisms which exist between trees, and also how they demonstrate 'learning'. He refers to the work of František Baluška of the Institute of Cellular and Molecular Botany at the University of Bonn who suggests that trees have brain-like structures at their root tips which emit measurable electrical impulses and contain 'signaling pathways' and 'molecules similar to those found in animals' (83). Wohlleben argues trees live in community, for example trees growing near each other form a single canopy (see Figure 4.1). Such a joint canopy protects the individual trees in storms and being in close proximity also enables a sharing both of sunlight and of nutrients via fungal networks connecting their roots.

Wohlleben acknowledges the scepticism of the majority of biologists that such features or behaviours of trees point towards trees having capacity for

Figure 4.1 A group of trees forming a single canopy
Source: [Author's photograph]

intelligence, memory and emotions and also notes the concern of many biologists about the potential breaking down of barriers between plant and animal
life. However, he then asks why such category keeping is so important to
Western scientists. As highlighted in Chapter 2, this focus on categorisation,

identified by Little Bear (2016) as a Western mindset towards the world, can lead to analysis based on separate parts rather than, as in many Indigenous systems, a focus on the whole and on interconnections (also see De Line 2016). Challenging the categories and boundaries between the human and the wider natural world is central for many contemporary Western thinkers such as Braidotti (2013b) and Haraway (2007, 2016) who foreground that humans are themselves composites of human, mineral and non-human-others. Proposing the prospect of subjectivity in plants and other parts of the wider natural world such as land is not to suggest that their subjectivity is the same as human subjectivity but could still be possible and meaningful.

A further problematic issue for dominant Western notions of the human subject is that as well as being anthropocentric (human centred) they are also androcentric (male centred) and Eurocentric as brought together in the Eurocentric conception of 'universal man'. In such a framing, the human subject is 'white, male, heterosexual, urbanized, able-bodied, speaking a standard language and taking charge of the women and the children' (Braidotti 2013b: 2). Such a Eurocentric universalised subject is visually represented in Leonardo da Vinci's Vitruvian man (see Figure 4.2). As Braidotti (2013b: 2) explains, Vitruvian man represents 'an ideal of bodily perfection which doubles up as a set of mental, discursive and spiritual values'. This iconic image is the emblem of Humanism – a Eurocentric (Western) doctrine that 'combines the biological, discursive and moral expansion of human capabilities into an idea of teleologically ordained, rational progress. Faith in the unique, self-regulating and intrinsically moral powers of human reason forms an integral part of this high-humanistic creed' (2).

'Eurocentrism' does not 'identify a geographical location or contingent attitudes' (2). Instead, 'it is a structural element of our cultural practice, which is also embedded in both theory and institutional and pedagogical practices' (2). As highlighted in earlier chapters, these practices extend across the world, their spread driven through colonialism and continuing colonial influences on structural elements of societies worldwide, including in education. There has, however, been an increasing questioning of this 'high-humanistic creed', as I have explored in the opening chapters of this book. As Braidotti (2013b: 4) points out: 'Sexualised, racialised and naturalised differences . . . have evolved into fully-fledged alternative models of the human subject'.

Thinking about subjectivity is also problematic in that it requires the subject to be capable of theorising about the nature of the subject, i.e. theorising about itself.[1] To use Foucault's phrase, in attempting to theorise subjectivity, man[woman] is both 'an object of knowledge and a subject that knows' (Foucault 1973: 386 cited in Biesta 2006: 3). Rather than accept this difficulty as insurmountable the thinkers Biesta (2006, 2010, 2013) and Braidotti (2011a, 2011b, 2013a, 2013b, 2016) provide other ways to approach the issue of theorising about subjectivity. Whilst their ideas are very different, I argue that both provide some very useful insights for education which seeks to be sustainable and democratic. From their thinking, and also reflecting on thinking explored in Chapters 2 and 3, I highlight ideas which I take forward and develop further.

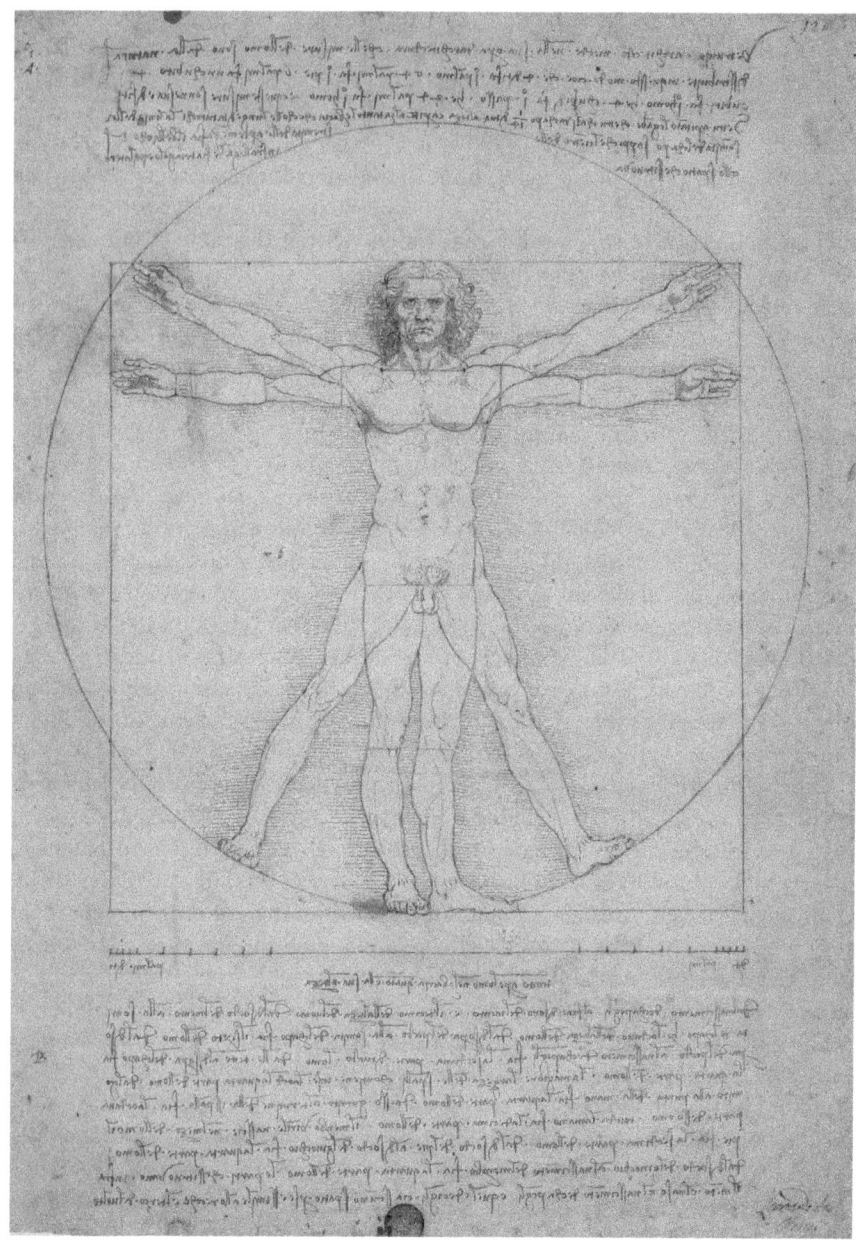

Figure 4.2 Vitruvian Man

[Pen and ink with wash over metal point on paper] (1490), Leonardo da Vinci, Gallerie dell'Accademia, Venice

Source: [Reproduced with permission of The Ministry of Cultural Heritage and Activities for Tourism, Galleries of the Academy of Venice]

Biesta's 'pedagogy of interruption'

Biesta (2006) suggests that rather than simply giving up theorising about what it is to be a human subject in the face of the difficulties outlined by thinkers such as Foucault (1973) (see also Derrida 1991), it is possible instead to reframe the question to explore '*who comes into presence?*' as a human subject and *when* this matters. He has developed these ideas in his 'pedagogy of interruption', which he proposes as a new theory of education. Biesta's 'pedagogy of interruption' (2010, 2013) can be understood as a *posthumanist* pedagogy. He draws on Foucault's analysis of the subversion of the modern conception of 'man' as well as Levinas' critique of humanism as not sufficiently human, as demonstrated by the inhuman acts witnessed in recent history such as in Stalinism, the Holocaust and continuing war and genocide. It is not, however, a pedagogy which 'comes after humanism or at the end of humanism' (Pedersen 2010: 242) in a chronological sense but rather one which opens up the possibility of moments which fleetingly '*interrupt*' the existing rational community.

Biesta identifies two key features – '*coming into the world*' and '*uniqueness*' in his thinking on what he calls the process of 'subjectification'. Drawing on the work of Jean-Luc Nancy, Biesta (2013: 143) initially explored the importance of the idea of 'coming into presence' which he found was 'a much more existential way to talk about the subject, one that referred to an *event* rather than an essence or identity, and one that expresses an interest in *who* comes into presence rather than in an essence or identity that tries to define "what is to come, ought to come, or is allowed to come into presence"'. For Biesta:

> the idea of coming into presence thus turned traditional educational thinking on its head by not starting from what the child is to become, but by articulating an interest in that which announces itself as a new beginning, as newness, as *natality* [my italics], to use Arendt's term.
>
> (143)

It was from engaging with the work of Hannah Arendt and also the architect Bernhard Tschumi, that Biesta then moved from an interest in 'coming into presence' to one of 'coming into the world'. For Arendt (1974 [1958]), coming into the world can never be in isolation, it must involve action in the presence of others. Moreover, Biesta (2013: 143) comments that:

> if we are committed to a world in which everyone's beginnings can come into presence, we have to live with the fact – which is actually not a fact but an articulation of what it means to exist politically (see Biesta 2010) – that the ways others take up my beginnings are radically beyond my control. The very condition that makes my 'coming into presence' possible – the ways in which others take up my beginnings – also disrupts the purity of my beginnings, so to speak, as others should have freedom to take up my beginnings in their own way. Arendt's phrase 'plurality is the condition of human action' still captures this very well.'

In this logic, subjectification – understood as 'coming into the world' – necessarily takes place in a public space of plurality and difference and is thus inherently political.

However, 'coming into the world' would not be enough on its own. For Biesta (2010, 2013), 'uniqueness' is also a key issue with regard to subjectivity. He comments that:

> If we were only to have 'coming into the world' we would have an account of how the event of subjectivity occurs – a theory of subjectivity to put it differently – but we would not have an argument for *why* the subjectivity of each single subject might matter.

To explore this issue of why an event in the process of becoming a human subject *might matter*, Biesta draws on the work of Levinas (2000 [1961]) and Lingis (1994). Biesta proposes that Levinas and Lingis understand uniqueness as *irreplaceability*.[2] Instead of asking '*how* am I different from others?' it asks instead '*when* does it matter that I am unique?' For Biesta, the answer to this is that it matters when *I* am being addressed specifically, when *I* am singled out rather than it being a case of 'I' as but one representative of the rational community, i.e. 'I' in my social role or identity such as teacher, doctor, child in year six. In a call where *I* am irreplaceable, such irreplaceability brings with it a responsibility for the other who is making such a call. Whether we take up this responsibility is entirely up to us. It is an ethical situation and also an existential one, based on an event. It claims 'nothing about what the subject is – just about the situations we find ourselves in' (2013: 145). Our uniqueness is revealed as an event rather than in an essence which an individual possesses, thus avoiding the 'impasse' created by attempts to theorise about the nature or essence of the subject.

In developing his thinking Biesta thus identifies two key ways to articulate uniqueness – uniqueness understood as *difference* (how we differ from other people, a third-party perspective) and uniqueness as *irreplaceability*. Biesta (2006: 85) argues that Arendt's understanding of uniqueness, unlike that of Levinas and Lingis, runs the risk of conceiving of uniqueness in terms of characteristics or qualities of the subject – and would thus conceive of 'unique' in terms of what we have or possess, commenting, 'It would, to put it differently, turn the question of uniqueness into a question of identity' (Biesta 2010: 85). Counter to Biesta, I argue, however, that Arendt is well aware of this danger. For example, she highlights in *The Human Condition* 1974 [1958]: 181) that the manifestation of '*who* a person' is curiously intangible, confounding verbal expression. She comments 'the very moment we want to say *who* someone is, our very vocabulary leads us astray into saying "*what* he is", characteristics one shares, (and does not share) with others'. Arendt proposes, however, that *who* a person is '*plainly visible*' in first-person encounters enacted in the web of relations with others in the world. I return to this issue in Chapter 7 when I discuss the possibility and value of first-person intersubjective encounters, including in education, and ways that such encounters can be encouraged, although never guaranteed.

To return now to Biesta's line of thinking – Biesta (2013) argues that subjectification understood as 'coming into the world in our uniqueness' does not fit into an understanding of education as a place of production – for example the production of certain learning outcomes or particular kinds of subjects. Instead, moments 'interrupt' the rational community (which does also have a valuable role and one which these moments are not trying to replace) as fleeting events. However, education can very easily prevent such events if there is no opportunity to be addressed by the other; to hear the call of the other; to 'open our ears and perhaps even our hearts' (Biesta 2013: 146).

Pedersen (2010: 243), however, points out that when Biesta critiques the modernist conception of 'man':

> he argues for an open-ended approach to human subjectivity without interrogating the position of 'the human' in education as such. In this sense, the category of 'the human' is still ascribed authority, with inquiries into the co-constitution of human–animal subjectivities and alterities closed off, and without conceptual space for dealing with the myriad ways in which nonhuman animal presences are always already part of 'our' human selves.

Pedersen (2010: 243) highlights how questioning this authority of 'the human' enables an expansion from Biesta's position, in which 'we can only come into presence in a world populated by other human beings who are not like us' (Biesta 2006: 32). This opens up possibilities for approaches to 'coming into presence', which allow for a much broader range of 'others unlike us'. Biesta does not exclude possibilities of encounters with other-than-humans, but his lack of acknowledgement and exploration of such encounters leaves unchallenged the authoritative, and currently exploitative, position of the human in relation to the wider natural world.

Braidotti's nomadic subject

Braidotti proposes a very different way to address the difficulties identified by late twentieth-century Western philosophers such as Foucault of theorising about subjectivity. Braidotti (2002, 2011a, 2011b, 2013b) argues for an 'embodied materialism' and an 'embodied or enfleshed subject' (2011b: 15) in which subjectivity is explored as 'a lived experience' rather than as an attempt to 'gaze' upon it from the outside. Within this framing, she has developed what she calls 'a process of becoming nomad' (Braidotti 2011b: 5). Here she is developing, through her engagement with the ideas of Deleuze, a line of thinking which can be linked back to Bergson (see Chapter 2) and his conception of *élan vital*. The figuration of 'nomad' emphasises *immanence*, arising in different situations including 'in relation to, and in interaction with, marginalised others' (5). The 'enfleshed' nomadic subject 'renders an image of the subject in terms of a non-unitary and multi-layered vision, as a dynamic and changing entity' (5): a subjectivity which 'zigzags' (2006). The nomadic subject challenges universalism and destabilises static, autonomous conceptions of the subject, particularly humanist understandings of the subject dominant in Western thinking (i.e. white, European, able-bodied, rational human males). Braidotti's emphasis on immanence resonates with the ideas of Gilligan (1982) and other feminist thinkers (introduced in Chapter 3) who emphasise immanence in their exploration of ethics. For Braidotti (2011b: 7), the point of nomadic theory is to:

> Identify lines of flight, that is to say, a creative alternative space of becoming that would fall not between the mobile/immobile, the resident/the foreigner distinction but within all these categories. The point is neither to dismiss or glorify the status of the marginal, alien others, but to find a more accurate, complex location for the very terms of their specification and our political interaction.

Braidotti (2011b: 15) argues for the importance of re-grounding oneself in order to critique power radically and to contribute to the dissolution of the humanist subject. To achieve this shift in perspective she draws on the 'feminist politics of location', a:

> method as well as a political tactic that aims at accounting for the diversity and complexity within any given category like women, feminists, lesbians, gays – while avoiding cognitive and moral relativism and thus safeguarding political and ethical agency.

This approach concentrates its methodological efforts on 'the analysis of the multiple power locations one inevitably inhabits as the site of one's subjectivity' (15). The possibility of such analysis, and subsequent 'undoings' of engrained power differentials, requires, for Braidotti, awareness of two crucial notions: memory and location. Such awareness is challenging since one often does not 'see' one's own location – it escapes 'self-scrutiny'. Feminist politics of location addresses this lack by using consciousness-raising to 'estrang[e] us from the familiar' (16), 'reterritoriali[sing] us, to use Deleuze's phrase' (16) and engaging us in micropolitical actions. The 'immanent regroundings' (15) which arise in and through this process take place in *embodied events and encounters* – a central theme for this book, as subsequent chapters explore.

Braidotti's concept of immanent, embodied, nomadic subjectivity also opens the possibility of *posthuman* subjectivity. For Braidotti (2016: 686) 'the relational capacity of the post-anthropocentric subject is not confined within our species but includes all non-anthropocentric elements: the nonhuman vital force of life'. Here one hears echoes of Bergson's (1912: 103) *élan vital* – the 'process of becoming' driven by 'the explosive force – due to an unstable balance of tendencies – which life bears within itself'.

Braidotti's vitalist approach to living matter:

> displaces the boundary between the portion of life – both organic and dis-
> cursive – that has traditionally been reserved for *anthropos*, that is to say *bios*,
> and the wider scope of animal and nonhuman life also known as *zoē*.
>
> (2013b: 60)

Braidotti recognises that her understanding of *zoē* is very different from the negative understanding proposed by Agamben (1998),[3] commenting that '[h]uman subjectivity in this complex field of forces has to be redefined as an expanded relational self engendered by the cumulative effect of all these factors' (Braidotti 2013b: 2).

The terms 'posthumanism' and 'posthuman' are often used interchangeably in academic and popular literature. I find that a useful 'working definition' is as follows. 'Posthumanism' relates to an understanding of the human not limited by Eurocentric notions of 'the human/humanism'. The posthuman, as in Braidotti's thinking, however, challenges the very boundary between the human and the wider natural world. As Pedersen (2010: 243) comments, Braidotti's thinking provides 'a conceptual space for dealing with the myriad ways in which nonhuman animal presences are always already part of "our" human selves'.

In the terms 'posthuman' and 'posthumanism' the prefix 'post' does not imply a chronological progression beyond humanism. Instead it opens a way to ask, within existing Western philosophical framings, fundamental ontological and epistemological questions of *what it is to be a (human) subject acting in the world*. Critics of posthumanism/the posthuman such as Williams (2018) comment that the current ecological crisis is not a time to 'give up being human', arguing that it is a time instead to take responsibility as humans for the ecological problems humans have created. However, this is a misunderstanding of the posthuman as proposed by Braidotti (2006). What is challenged in Braidotti's posthuman framing are the boundaries between humans and others in the world and the anthropocentric and androcentric superiority attributed to humans in Western (Eurocentric) thinking.

Braidotti's exploration of vitalism and posthuman intra-connection resonates with the various thinking on flux and the Indigenous *all my/our relations* introduced in Chapters 2 and 3. In developing her line of thinking, Braidotti is reaching back through Deleuze and Bergson to the ideas of Spinoza rather than appropriating the ideas of Indigenous peoples (she does show awareness of the thinking of Indigenous people, for example see Braidotti 2012). It is important, however, that others, when taking up her ideas, do not create a new form of universalisation of Western ideas and a denial that other ways of thinking have long existed: a danger highlighted for example by Todd (2016) in her critique of the Western Academy's posthuman and posthumanist turn.

Through Deleuze, Braidotti also reaches back to Spinoza including his thinking on *potentas* and *potentia*. Spinoza, a complex seventeenth-century thinker, was responding to the many intellectual developments in the era in which he lived. Briefly put, he emphasised intertwining between all of nature and challenged mind-body dualism, proposing that both emanated from the same substance (see Nadler 2020, Steinberg 2019 for further discussion). From this overall starting point, he developed a number of ideas, including his thinking on where power is located. For Spinoza, *potentas* is a sense that power is held by others whilst *potentia* is a sense that power is fluid, dynamic and immanent, residing within one and that individuals have a power potential to effect change. Nomadic subjectivity and engaging in feminist politics of location emphasising self-scrutiny/consciousness-raising and micropolitical action are expressions of *potentia*: an immanent power potential for action and changes in *this world*. Braidotti (2012) draws attention to the potential and positivity of action as opposed to the despair so often prominent in nineteenth- and twentieth-century philosophical writing. Braidotti notes such positivity in the writings of Hannah Arendt alongside Arendt's focus on action and on love for the world. Braidotti (2019: online) highlights how it is conceiving of the self in relational rather than individual terms, brought together with the power to act, which opens the possibility of positivity and hope. She foregrounds how '[t]he ethical life is the pursuit of relations, situations, contexts and values that enhance our power to act in the world'. It is about our power 'to take in the world's pain and process it'. In addition, it is about the ethical responsibility she feels as a teacher and researcher 'to work for hope and on the construction of a "we" that can take on this task and work in the direction of joyful, gratuitous experimentation regarding what we are capable of becoming' (online).

The posthuman thinking discussed here is also developed in Haraway's (2007, 2016) writing on 'companion species', which she understands as those we gaze upon in close encounter.[4] In such relationships we are 'beings-in-encounter' (5), where one is created in inter- and intra-action in which:

> the partners do not precede the meeting; species of all kinds, living or not, are consequent on a subject-and-object-shaping dance of encounters.
>
> (4)

Haraway (2000: 55) argues that:

> No objects, spaces, or bodies are sacred in themselves; any component can be interfaced with any other if the proper standard, the proper code, can be constructed for processing signals in a common language.

Building on the ideas of both Haraway and Braidotti, Ferrante and Palmieri (2015) argue for approaches in education which encourage opening spaces of 'intermingling' *in the present*: spaces where taking care of one's own becomings and the intra-connected becomings of others is possible. These are spaces where playfulness with many and varied ideas is possible, a playfulness generating complex interactions which can open spaces which do not attempt to colonise the future with these current understandings and ways of being but instead leave it radically open for the emergence of new subjectivities and new futures.

Biesta and Braidotti: taking their ideas forward

Braidotti and Biesta explore the issue of subjectivity in very different ways. Biesta emphasises the problematic nature of theorising about the subject and chooses to reframe the question as one of *who* appears and *when* does this matter. He emphasises uniqueness as irreplaceability. Braidotti does theorise about the subject, but it is a subject where the mind and body are not separate but instead embodied: experiencing and responding to the world as one. She argues for a nomadic subjectivity and the possibility of identifying: 'lines of flight, that is to say, a creative alternative space of becoming' (2011b: 7). However, there are aspects common to both thinkers and it is these aspects I take forward and develop in this book.

The first aspect is the importance of *an existential first-person event or encounter with the other*, encounters which emphasise *proximity*. In Chapters 7 and 8, I argue that such first-person/first-being encounters can be intersubjective encounters – encounters in which who one is as a unique being emerges in and through the encounter with the other. In this I am departing from Biesta who builds his argument around asking who emerges and when it matters (which he argues avoids the need to theorise the subject *per se*). Biesta (2013) does acknowledge that his 'pedagogy of interruption' is a starting point for further engagement and experimentation by other educators, and the ways that his ideas will then be taken up are not within his control. I acknowledge how Biesta's pedagogy of interruption has contributed to my thinking but also identify that my thinking on intersubjective encounters is different from his approach.

A second feature emphasised by Biesta and Braidotti is that encounters necessarily occur in and through *plurality*. This is when one encounters others unlike oneself, and when each is open to the expression of the other. This is a challenging process due, for example, to often-hidden dominant worldviews as discussed in the opening chapters.

Third, Biesta and Braidotti both value aspects of the thinking of Hannah Arendt. Braidotti (2012) recognises Arendt's concern for love of the world; her valuing of action in this world; the role of *potentia* in Arendt's understanding of power and the potentiality (and hope) this can open for the emergence of new futures. Biesta, amongst other things, acknowledges Arendt's thinking on *natality*. He highlights Arendt's concern for the possibility for newcomers to insert new ways of being (new subjectivities) into the world; how this is an expression of freedom and, as such, is an important democratic move.

Reflecting on indigenous thinking on subjectivity introduced in Chapters 2 and 3

Thus far in this chapter I have focused on ideas of human subjectivity in dominant Western thinking; the difficulties inherent in theorising subjectivity and the ways that two very different thinkers, Biesta and Braidotti have addressed such issues. I have identified themes which emerge for me and which I take forward to explore further. It seems necessary at this point also to reflect on the thinking explored in Chapters 2 and 3 on other framings of the world and subjectivity. My engagement with these ideas acknowledges that many framings of the world are possible and indeed valuable but that these often go unnoticed, unexamined in dominant cultures, limiting the possibilities for complex interactions which can enable new ways to emerge.

In reflecting on engagement with Indigenous ways of understanding and being in the world, Rosiek *et al.* (2020) highlight how one can easily slip into 'cherry picking' – 'Eurocentric scholars devolving into a form of settler colonial appropriation – taking a few things they feel are useful – but failing to commit to solidarity in broader projects that bring amelioration to Indigenous communities'. As highlighted in Chapter 3, they also go on to comment that such fear of 'cherry picking' can hold back possibilities of 'respectful' engagement, thereby creating a new form of silencing. Collaboration is one way forward, although this places a burden on Indigenous scholars to support Western scholars in this way, with what Tuck (Smith *et al.* 2018: 15) highlights are often very disappointing results in terms of improving conditions and correcting injustices. To respond to these difficulties, I am resisting highlighting particular aspects from the thinking shared in Chapters 2 and 3. Instead I would encourage the reader to engage directly with Indigenous and other thinking which resists dominant Western conceptions of subjectivity, drawing on Chapters 2 and 3 as starting points. I do return to ideas from Indigenous thinking, particularly when I explore intersubjective encounters in Chapters 7 and 8.

I would also reiterate here that what is explored in this book is *not* recommendations for a particular subjectivity or the 'correct' framings of the world. Instead, what is argued for is an engagement with a wide range of existing framings, including respectful reflection on commensurabilities and incommensurabilities, and the opening of spaces in and through which new ways can appear. For example, Rosiek *et al.* (2020) consider and value what can emerge in respectful engagement between Western new materialist thinkers and the Indigenous tenets identified by Deloria (1999) and Watts (2013), including both commensurabilities and incommensurabilities.

Opening spaces for the possibility of such emergence is central to the thinking of Hannah Arendt who emphasises what she calls 'spaces of appearance' and the potential for 'natality', and it is therefore to her ideas I now turn.

Hannah Arendt and opening 'spaces of appearance'

Arendt reminds us that 'Men not man live on the earth' (1974 [1958]: 176), and each individual person is 'unique, un-exchangeable and unrepeatable' (97). It is through acting and speaking in the presence of others, who are themselves unique beings, that one discloses one's uniqueness. However, this is not a disclosure in the sense of disclosing an already formed 'inner essence'. *Who* one is emerges in and through speech and action with others. It is in the 'space between I and we' (Topolski 2015: 176) that 'who I am' emerges. Arendt calls spaces where such disclosure/emergence is possible *'spaces of appearance'*.[5] Such spaces *open up in and through the **polis***, to use the term employed by the Ancient Greeks, although Arendt emphasises the use of the term in a particular way, commenting:

> properly speaking, it is not the city state in its physical appearance; it is the organisation of the people as it arises out of acting and speaking together, and its true space lies between people living together for this purpose, no matter where they happen to be. 'Wherever you go you will be a *polis*'.
> (Arendt 1974 [1958]: 198–199)

For Arendt, the term *polis* expresses:

> the conviction that action and speech create a space between the participants which can find its proper location almost any time and anywhere. It is the *space of appearance* [my italics] in the widest sense of the word, namely the space where I appear to others as others appear to me, where men exist not merely like other living or inanimate things but make their appearance explicitly.
> (199)[6]

For Arendt, whenever people gather, *a space of appearance* 'is potentially there, but only potentially, not necessarily, and not forever' (199). It is important to note here that Arendt, whilst drawing on Ancient Greek conceptions, is writing from the richness of the 'borderlands'[7] between Hebraic and Western thinking, bringing different ways of thinking and being into conversation and generating new ways to think about the world.

Speech and action take place in the public sphere or realm. Action is the highest category in what Arendt calls the *vita activa*.[8] The other two categories of the *vita activa* are labour (the activity needed to maintain us physically) and work[9] (which creates relatively permanent artefacts in the world). Labour and work take place in what she calls the private sphere or realm. Arendt values these activities but what concerns her is the rise of an attitude in which everything is treated *as though* it is part of an endless cycle of labour and the consumption needed for bodily functioning and survival. She states:

> In our need for more and more rapid replacement of the worldly things around us, we can no longer afford to use them, to respect and preserve

their inherent durability; we must consume, devour, as it were, our houses and furniture and cars as though they were the 'good things' of nature which spoil uselessly if they are not drawn swiftly into the never-ending cycle of man's metabolism with nature.

(1974 [1958]: 125–126)

Arendt (1996 [1929], 1974 [1958]) argues that this does not lead to fulfilment: instead it leads to a fear of loss, or further unfulfilled desire. Moreover, for Arendt, a focus on labour, work and cycles of consumption hijacks the political realm as a place to pursue one's own private physical needs (what Arendt terms the 'rise of the social') 'rather than being a venue for real political action' (Voice 2014: 42). This results in a situation where individuals see themselves as *consumers* of politics rather than active participants in the political process – participants who can *enact change* and facilitate, for example and importantly for this book, the emergence of new and more sustainable ways of being and living in the world. The 'rise of the social' has ethical implications. As Arendt comments in *The Origins of Totalitarianism: Part Three* (1968 [1951]: 36), 'Nothing proved easier to destroy than the privacy and private morality of people who thought of nothing but safeguarding their private lives'.

For Arendt, opportunity *to speak and act in the public realm* **is** *freedom*, since it allows a person to emerge as a unique subject, and thus is a democratic move. It is the denial of such a public space in which to speak and act that is the hallmark of totalitarian regimes. It is in speech and action in the public realm that '*who*' one is, not '*what*' one is, emerges. This distinction between *who* a person is and *what* a person is is crucial in Arendt's work. '*What* one is' is about a person as a representative or example of a species, a social group or even an educational age group – a child in year six, an adult learner: more or less physically, biologically the same as others in the group. '*Who* one is', in contrast, is about an individual in their *uniqueness*.

Appearance, for Arendt, also has another key feature – what she calls '**natality**'. Natality is integral to (a part of) appearance rather than something that happens within the space. It is these concepts – 'spaces of appearance' and 'natality' – held together as one, which make Arendt's philosophy both 'ruptural and inaugurative' (Dikeç 2013: 78) and 'complexity-compatible'.[10] Natality emphasises that we are born and continue to have new beginnings in life *in this world* rather than an emphasis on death, as, for example, in Heidegger's 'being unto death',[11] or a focus on the afterlife as in Christian doctrine. Moreover, rather than a focus on an 'isolated self' as experienced in death, natality involves a creative act 'between plural selves', between 'a we rather than an I' (Hayden 2014: 15). Arendt's thinking on 'natality' began in her thesis on *Love and St Augustine* and finds a fuller expression in *The Human Condition*. In *Love and St Augustine*, Arendt (1996 [1929]: 46) began to explore Augustine's notion of *initium*:

> *Initium ergo ut esset, creatus est homo, ante quem nullus fuit* ('That there be a beginning, man was created before whom there was nobody' [Arendt's translation]).

In her interpretation of *initium* Arendt sees that the possibility for immortality is to be found in acting, being an *initium* in this life, this world, rather than in the afterlife as in the thinking of St Augustine. In *The Human Condition* (1974 [1958]: 177), Arendt develops her conception of *initium*, commenting how 'beginning' in the sense of an *initium* 'is not the same as the beginning of the world, it is not the beginning of something but of somebody, who is a beginner himself'. The event of birth of an *initium* into the world (Arendt's 'natality') has a sense of both birth and 'second birth'. For example, she says:

> With word and deed we insert ourselves into the human world, and this insertion is like a second birth, in which we confirm and take upon ourselves the naked fact of our original physical appearance.
>
> (1974 [1958]: 176–177)

For Arendt, action and novelty are intrinsically linked to natality since:

> To act, in its most general sense, means to take an initiative, to begin. . . . Because they are *initium*, newcomers and beginners by virtue of birth, men are prompted into action. . . . It is the nature of beginning that something new is started which cannot be expected from whatever may have happened before. This startling unexpectedness is inherent in all beginnings and all originals.
>
> (1974 [1958]: 177)

Such novelty opens the possibility of disruption of existing dominant Western Enlightenment thinking which frames subjectivity in terms of rationality and autonomy, including in the field of education. Beginnings, however, do raise important questions about ethics since our beginnings are 'unbounded' (cannot be 'bounded' by us, as others are involved who may take things in directions we did not anticipate); 'irreversible' (once brought into the world beginnings cannot be taken back) and unexpected. If education encourages new ways of being and doing to emerge this raises the question of what happens if some do not feel that these ways are ethical or compatible with differing conceptions of what is understood as 'good'. I am *not* proposing that because something is new it is 'good'. However, as Biesta (2006) points out, what is important in education is that new ways have *the possibility* of emerging, even if they are then challenged, interrupted. In *The Human Condition* (1974 [1958]), Arendt proposes forgiveness and mutual promising as an ethical response to the unexpectedness, unboundedness and irreversibility which emergence of the new generates. I explore these ideas in Chapter 6, where I engage with the challenging issue of ethics within the logic of emergence.

Arendt's ideas on natality and on opening 'spaces of appearance' – spaces to speak and act with others as an expression of freedom – have a 'world building' focus and potential to encourage sustainable and democratic education.

However, drawing Arendt's ideas into educational settings is complex and controversial for reasons I now go on to explore.

'Spaces of appearance' in education: issues and possibilities

In Arendt's conception, spaces of appearance have the potential to occur 'where I appear to others as others appear to me' (1974 [1958]: 198). At first sight the possibility of such spaces occurring in education seems straightforward. However, for Arendt such spaces appear when I act, speak and begin something new in the *public sphere/domain*. Whether and how education can or should be considered such a 'public' space has caused much debate in academic literature. Since encouraging spaces of appearance in education is central to the arguments in this book, I now examine these issues in some detail, including discussion of Arendt's essay *Crisis in Education* (2006a [1961]) and her controversial writing on American school integration in *Reflections on Little Rock* (1959).

Arendt has been represented in the literature (for example see Biesta's earlier writing, 1996)[12] as positioning education in the private domain. However, in *Crisis in Education* (2006a [1961]: 185) Arendt clearly states:

> Normally the child is first introduced to the world in school. Now the school is by no means the world and must not pretend to be; it is rather the institution that we interpose between the private domain of the home and the world in order to make the transition to the world possible at all.

Some thinkers, for example Biesta (2013), argue that it is not possible to hold separate the realms of politics (the public domain) and the school. They contend that such separation is based on seeing education in physiological terms: a developmental understanding of the child who at some point becomes an adult who then enters the public domain of (political) speech and action.[13] In addition, such thinkers argue that separation of school from the public domain is based on a false assumption that 'the dynamic realm of politics – the dynamics of beginning and responding, of action in plurality – either do not happen in the realm of education or can be held at bay by the educator' (Biesta 2013: 112). In contrast, other thinkers (for example see Timmermann Korsgaard 2016) argue that separation between education and the (political) public realm is vital, supporting a key point in Arendt's argument – that education should be a place of *protection*:

> for the free development of characteristic qualities and talents . . . the uniqueness that distinguishes every human being from every other, the quality by virtue of which he is not only a stranger in the world but something that has never been here before.
>
> (Arendt 2006a [1961]: 185)

In this line of thinking it is important to protect the school from a 'politicisation' which would change the very nature of the school as *a place for the safe study of the world*. Such a 'protected' space allows students to play with abundant materials and possibilities, opening potential for the emergence of new ways of being in and knowing the world. Understanding education as a place of protection is close to the original sense of the Ancient Greek word *skholé* which is the etymological root of the word 'school'. Literally translated *skholé* means free time, a break, a respite, leisure. However, for the Ancient Greeks, this is not leisure in the sense of a luxury or break from a primary activity. Rather it denotes a time which has a higher value than what it is interrupting. It is a time to debate, to reflect, to ponder new, unexpected possibilities. This is not to say that issues of the world should not be explored in education understood in this sense of *skholé*. However, what the approach does highlight is that children should be protected from the responsibility of solving the political issues of the day (be that, for example, school integration, knife crime or environmental degradation). In this line of thinking, education as *skholé* can be:

> a place of experimentation with the possibility of the impossible precisely because education does not have to make political decisions about the future. . . . This does not mean dropping all prior values and throwing away the lessons of the past. It means, rather, using the lessons of the past to invent something radically new; something which might accompany us into the future (and also which might not).
>
> (Osberg 2010: 164)

Timmermann Korsgaard (2016) draws on Arendt's *Reflections on Little Rock* (1959) in which she discusses the highly political events which occurred in the American city of Little Rock, Arkansas in 1957 to illustrate this issue of protection. Following the Federal decision three years earlier that separating educational establishments on the basis of race was inherently unequal and therefore unlawful, NAACP[14] activists in Little Rock recruited nine children prepared to enrol in the town's all-white high school. On the first day of term an angry mob and the local guard prevented the entry of these youngsters into the school. For all the students this was a traumatic experience, but for one student, Elizabeth Eckford, the experience was particularly distressing (Margolick 2011). Foreseeing trouble, the local organisers had telephoned the children's homes at the last minute, asking that the children should be accompanied, planning to enter the school as a group. Elizabeth and her parents, having no telephone, were unaware of this. Elizabeth unsuccessfully attempted the entry alone and finally was helped to walk away by a friend of her father. Arendt (1959: 50) comments that the 'girl, obviously, was asked to be a hero' and adds that the white children captured in photographs (still widely available today) will also find it hard to 'live down' the brutality they displayed that day when they may have moved on from such a brutal stance in later life.

Reflections on Little Rock was very controversial, and its publication was delayed for over a year. In the article Arendt emphasises that children should not be politicised and used to fight the battles which adults have confessed themselves unable to resolve. She argues that the law which, at that time, prevented marriage between black and white citizens was an issue of human freedom and the human right to a private life. She felt that this is where political action should have been focused initially rather than on forcing integration into schooling. hooks (1994: 4) writes about her own and others' difficult and often damaging experience of having to attend white schools as part of the US school integration strategy. She describes her shock at experiencing school 'no longer as a place of ecstasy' but instead as a place where we were 'always having to counter white racist assumptions that we were genetically inferior, never as capable as white peers, even unable to learn'. School was no longer a place opening possibilities for emergence of new futures and new ways to be in it. hooks comments:

> Those periods in our adolescent lives of racial desegregation had been full of hostility, rage, conflict and loss. We black kids had been angry that we had to leave our beloved all-black high school, Crispus Attucks, and be bussed halfway cross town to integrate white schools. We had to make the journey and thus bear the responsibility of making desegregation a reality. We had to give up the familiar and enter a world which seemed cold and strange, not our world, not our school. We were certainly on the margin, no longer at the centre, and it hurt. It was such an unhappy time.
>
> (24)

Consideration of the way the word *political* is used introduces a way for *rapprochement* between the arguments for and against education as a place of protection from the political. As Masschelein (2011) and Masschelein and Simons (2013) argue, we can understand 'the political' in terms of *politicisation* of education. This would make education a setting where, through control of funding, curricula and educators, current political issues can be played or even fought out, where existing ideas of the future can be aimed towards and where children are used as *tools* to make particular futures happen or are expected *to take responsibility* for changes needed in society. Such politicisation works against the concept of *skholé* as a break, a space apart and could indeed be a barrier to the possibility of emergence of the new in education. However, if the 'political' is understood as arising in and through acting and speaking where each is open to the other, a place where natality is possible, then education *can* be 'political' in this sense.

Thus, to be consistent with *skholé* as a break, a place between the private and public spheres, education needs to be a place to break free from the political already existing in the world, including in the children's situated life[15] outside school.[16] Such educational spaces need an element of *protection* to ensure that they do not become political in the sense of 'politicised'.[17] Echoing Macmurray's proposal (introduced in Chapter 3) that sometimes problems emerge which make it challenging or impossible to go on as we are and it is therefore necessary to withdraw from action for a period of time to allow emergence of novel ways to respond, education can be a 'protected' place in which such a process is possible. In such democratic spaces (democratic in the sense they open freedom to be and act in unexpected ways) students can speak but they need opportunities to speak with their own voice rather than as mouthpieces or tools of the adult world.

I have argued in this chapter that it is valid and valuable to speak of 'spaces of appearance' in education understood as *skholé*. In such spaces, which open up through acting and speaking in the presence of others who are themselves unique beings, *who* one is as an *initium*, a beginner, can appear, opening possibilities for the emergence of new subjectivities and new futures. This does then raise the question of what could *encourage* the possibility of such spaces. Whilst accepting that 'Whenever people gather [a space of appearance] is potentially there, but only potentially, not necessarily, and not forever' (Arendt 1974 [1958]: 199), one might wish *to encourage* the possibility of opening such spaces. This is the focus of the next chapter. I continue to discuss Arendt's ideas and also draw on Andreotti, Mouffe, Rancière and Masschelein and Simons. I explore why it is helpful, despite the challenges it presents, to read these different thinkers together and consider how their ideas can contribute to sustainable and democratic education and the emergence of new ways of being together in our shared planet.

Notes

1 Other twentieth-century Western philosophers also critiqued Western notions of subjectivity. For example, Derrida highlights the need and responsibility to be 'rebellious toward the traditional category of "subject"' (Derrida 1991: 109). For further reading, see the collection *Who Comes after the Subject?* edited by Cadava *et al.* (1991).
2 See Biesta 2006, 2010, 2013 for further discussion of Levinas and Lingis.
3 In his thinking Agamben valorises *bios* – (discursive life of the citizen) over *zoë* – bare life.
4 'Species' has its etymological root in the word 'to gaze'. 'Companion' has its etymological root in those who are our *co-pain* – those with whom we 'break bread'.
5 Braidotti (2011b: 7) does not use the terminology of 'spaces of appearance' but highlights in her nomadic subjectivity the importance of identifying 'lines of flight' that which can open a 'creative alternative space of becoming'.
6 Arendt limits her thinking to the human realm: remember that Arendt was a thinker who died in 1975, so she cannot be expected to engage with more recent posthuman writing, but this does not prevent us from bringing her writing into conversation with later postmodern and posthuman ideas. In Chapter 8, drawing on recent posthumanist and posthuman thinking, I develop arguments which draw on the 'threads' of Arendt's conception of 'spaces of appearance' without allowing such threads to then be a 'chain fettering us', to use Arendt's phrase (2006b [1961]: 94), a move which I would like to think Arendt invites. I discuss and take forward the possibility of spaces of appearance in engagements with the wider natural world.
7 A term I introduced in Chapter 2 – see discussions in Anzaldúa (1999) and Mignolo and Tlostanova (2006).

8 Arendt's *vita activa* has two distinctive features which set it apart from other ways of theorising. First, it concentrates on, and validates, the active life and has a worldly focus. This stands in contrast to the *vita contemplativa* which has usually been privileged over the *vita activa* in Greek and Christian thinking. Second, it opposes the Western Modernist viewpoint which, whilst also rejecting a prioritisation of the *vita contemplativa*, either privileges labour and work (as seen for example in the writings of Marx) over the political sphere, or, as in the philosophy of Kant, which 'whilst articulating a philosophy of practical life, retains its foundation in universal reason and takes as its source of authority the noumenal realm which stands outside the practical' (Voice 2014: 44).

9 Labour is repetitive and leaves no permanent artefact or symbols in the world. It is cyclical in nature, unremitting and concerned with the production and consumption needed for biological functioning and survival. Arendt is not denigrating the necessity of labour, although some critics do accuse her of this (for example some feminists question her attitude to the care needed for biological functioning – see Dietz 1994).

Work is the product of *homo fabricans*. It differs from labour, as it leaves something (relatively) permanent in the world. Such objects, which stand outside humans' subjectivity create a world which stands between humans and can be recognised by them. Such recognition generates a stabilising effect on the common world. Modern life, with its rise in the disposable aspects of consumerism, threatens this 'objective world' since 'objects lose their use character and become more and more objects of consumption' (Arendt 1974 [1958]: 125).

10 In Chapter 2 I introduced the phrase 'complexity compatible', which Osberg (2015) uses to describe theories and ideas which whilst not identified as complexity theory are compatible with it and sometimes share some of its historical sources.

11 Katago (2014) suggests that whilst Heidegger does also explore 'being in the world' and notions of care (*sorge*), these are more limited since they are in relation to the individual self. Arendt's focus is on plurality in the world and thus her analysis is far richer.

12 For example, Biesta (1996: 97) comments 'While Arendt locates the interaction between adults in, by definition, the undetermined and undeterminable public realm of politics, she situates education in the private realm and makes a firm plea for separation between these two domains'. In his more recent work, Biesta recognises that Arendt locates education as *a place between* the public and the private.

13 Biesta (2013: 111–112) states he does have some sympathy for Arendt's point that children need space to develop their own new beginnings.

14 The National Association for the Advancement of Coloured People.

15 In the next chapter I recognise and discuss some of the issues around suspension of children's situated lives and the sensitivities teachers need to bring to this.

16 Also, see discussion in Biesta and Lawy (2006).

17 Biesta's understanding does, perhaps, imply this, but his emphasis on developmental issues in Arendt's argument underplays this aspect.

References

Agamben, G. (1998) *Homo Sacer: Sovereign power and bare life*. Heller-Roazen, D. (tr.). Stanford: Stanford University Press.

Anzaldúa, G.E. (1999) *Borderlands/La frontera: The new mestiza*. San Francisco: Aunt Lute Books.

Arendt, H. (1959) Reflections on little rock. *Dissent*, 53: 45–56. Available at: http://learningspaces.org/forgotten/little_rock1.pdf [Accessed 20.5.2015].

Arendt, H. (1974 [1958]) *The human condition*. Chicago, IL: University of Chicago Press.

Arendt, H. (1996 [1929]) *Love and St Augustine*. Vecchiarelli, J. and Stark, J. (eds.). Chicago, IL: Chicago University Press.

Arendt, H. (2006a [1961]) Crisis in education. In *Between past and future: Eight exercises in political thought*. London: Penguin.

Arendt, H. (2006b [1961]) What is authority? In *Between past and future: Eight exercises in political thought*. London: Penguin.

Bergson, H. (1912) *Creative evolution*. Mitchell, A. (tr.). London: Palgrave Macmillan.

Biesta, G. (1996) Education not initiation. *Philosophy of Education*. Available at: ojs.ed.uiuc. edu/index.php/pes/article/download/2247/942 [Accessed 10.9.2015].

Biesta, G. (2006) *Beyond learning: Democratic education for a human future*. Boulder, CO: Paradigm Publishers.

Biesta, G. (2010) *Good education in an age of measurement: Ethics, politics, democracy*. London: Paradigm Publishers.

Biesta, G. (2013) *The beautiful risk of education*. London: Paradigm Publishers.

Biesta, G. and Lawy, R. (2006) From teaching citizenship to learning democracy: Overcoming individualism in research, policy and practice. *Cambridge Journal of Education*, 36(1): 63–79. doi: 10.1080/03057640500490981

Braidotti, R. (2002) *Metamorphoses: Towards a feminist theory of becoming*. Cambridge: Polity Press.

Braidotti, R. (2006) The ethics of becoming-imperceptible. In C. Boundas (ed.) *Deleuze and philosophy*. Edinburgh: Edinburgh Press, pp. 133–159.

Braidotti, R. (2011a) *Nomadic theory: The portable Rosi Braidotti*. New York: Columbia University Press.

Braidotti, R. (2011b) *Nomadic subjects: Embodiment and sexual difference in contemporary feminist theory*, 2nd Edition. New York: Columbia University Press.

Braidotti, R. (2012) *Rethinking the human sciences – panel I*, 29th May. CLS Columbia. Available at: www.youtube.com/watch?v=UDdfAJ25A4w [Accessed 10.10.2018].

Braidotti, R. (2013a). Posthuman humanities. *European Educational Research Journal*, 12(1): 1–19. doi: 10.2304/eerj.2013.12.1.1

Braidotti, R. (2013b) *The posthuman*. Boston, MA and Cambridge: Polity Press.

Braidotti, R. (2016) Posthuman feminist theory. In L. Disch and M. Hawkesworth (eds.) *The Oxford handbook of feminist theory*. Oxford: Oxford University Press, pp. 673–698.

Braidotti, R. (2019) *Posthuman knowledge*. Harvard University Graduate School of Design, 12th March. Available at: www.gsd.harvard.edu/event/rosi-braidotti/ [Accessed 5.4.2020].

Carruthers, P. (1998) Natural theories of consciousness. *European Journal of Philosophy*, 6(2): 203–222. doi: 10.1111/1468-0378.00058

De Line, S. (2016) All my/our relations: Can posthumanism be decolonized? *Open! Platform for Art, Culture & the Public Domain*. Available at: www.onlineopen.org/all-my-our-relations [Accessed 23.1.2018].

Deloria, V. (1999) *Spirit and reason: The Vine Deloria Jnr. Reader*. Golden, CO: Fulcrum.

Derrida, D. (1991) 'Eating well' or the calculation of the subject: An interview with Jacques Derrida. In E. Cadava, P. Connor and J. Nancy (eds.) *Who comes after the subject?* New York: Routledge, pp. 96–119.

Dietz, M. (1994) Hannah Arendt and feminist politics. In L. Hinchmanm and S. Hinchman (eds.) *Hannah Arendt: Critical essays*. Albany, NY: SUNY Press, pp. 231–260.

Dikeç, M. (2013) Beginners and equals: Political subjectivity in Arendt and Rancière. *Transactions of the Institute of British Geographers*. Royal Geographical Society (with the Institute of British Geographers), 38(1): 78–90.

Ferrante, A. and Palmieri, C. (2015) Take care of becoming. Towards a post-humanist educational milieu. *Mondi educativi. Themi indagini suggestioni (Educational Worlds: Themes,*

Investigations, Suggestions), 5(1): 1–19. Available at: www.metis.progedit.com/anno-v-numero-1-062015-leducazione-ai-tempi-della-crisi/128-saggi/671-aver-cura-del-dive nire-verso-un-milieu-educativo-post-umanista-1.html [Accessed 22.9.2016].

Foucault, M. (1973) *The order of things. An archaeology of the human sciences*. New York: Vintage/Random House.

Gilligan, C. (1982) *In a different voice: Psychological theory and women's development*. Cambridge, MA: Harvard University Press.

Haraway, D. (2000) A manifesto for cyborgs: Science, technology and socialist feminism in the 1980s. In F. Hovenden, L. Janes, G. Kirkup and K. Woodward (eds.) *The gendered cyborg: A reader* London: Routledge, pp. 50–57.

Haraway, D. (2007) *When species meet*. Minneapolis and London: University of Minnesota Press.

Haraway, D. (2016) *Staying with the trouble: Making kin in the Chthulucene*. Durham, NC: Duke University Press.

Hayden, P. (2014) Illuminating Hannah Arendt. In P. Hayden (ed.) *Hannah Arendt: Key concepts*. Durham, NC: Acumen, pp. 1–22.

hooks, b. (1994) *Teaching to transgress: Education as the practice of freedom*. New York: Routledge.

Katago, S. (2014) Hannah Arendt on the world. In P. Hayden (ed.) *Hannah Arendt: Key concepts*. Durham, NC: Acumen, pp. 52–65.

Levinas, E. (2000 [1961]) *Totality and infinity. An essay on exteriority*. Lingis, A. (tr.). Pittsburgh, PA: Duquesne University Press.

Lingis, A. (1994) *The community of those who have nothing in common*. Bloomington: Indiana University Press.

Little Bear, L. (2016) *Blackfoot metaphysics 'waiting in the wings'*. Congress of the Humanities and Social Sciences Big Thinking Lecture, 1st June. Available at: www.youtube.com/watch?v=o_txPA8CiA4 [Accessed 18.11.2018].

Lyvers, M. (1999) Who has subjectivity? *Psyche: An Interdisciplinary Journal of Research on Consciousness*. Available at: http://epublications.bond.edu.au/hss_pubs/12 [Accessed 3.2.2016].

Margolick, D. (2011) *Elizabeth and Hazel: Two women of little rock*. New Haven: Yale University Press.

Masschelein, J. (2011) *Experimentum scholae*: The world once more . . . but not (yet) finished. *Studies in the Philosophy of Education*, 30(5): 529–535. doi: 10.1007/s11217-011-9257-4.

Masschelein, J. and Simons, M. (2013) *In defence of schools: A public issue*. Leuven: E-ducation, Culture and Society Publishers. Available at: http://ppw.kuleuven.be/ecs/les/in-defence-of-the-school/masschelein-maarten-simons-in-defence-of-the.html [Accessed 3.2.2016].

Mignolo, W.D. and Tlostanova, M.V. (2006) Theorizing from the borders: Shifting to geo- and body-politics of knowledge. *European Journal of Social Theory*, 9(2): 205–221.

Nadler, S. (2020) Baruch Spinoza. In E. Zalta (ed.) *The Stanford encyclopedia of philosophy*, Summer 2020 Edition. Available at: https://plato.stanford.edu/archives/sum2020/entries/spinoza/ [Accessed 6.7.2020].

Osberg, D. (2010) Taking care of the future. *Complexity theory and the politics of education*. Rotterdam: Sense Publishers, pp. 153–166.

Osberg, D. (2015) Learning, complexity and emergent (irreversible) change. In E. Hargreaves (ed.) *Sage handbook of learning*. London: Sage, pp. 23–40.

Pedersen, H. (2010) Is 'the posthuman' educable? On the convergence of educational philosophy, animal studies, and posthumanist theory. *Discourse: Studies in the cultural politics of education*, 31(2): 237–250. doi: 10.1080/01596301003679750

Pelt, J. (in collaboration with Steffan, F.) (2004) *La Solidarité chez les plantes, les animaux, les humains*. Paris: Fayard.

Rosiek, J., Snyder, J. and Pratt, S. (2020) The new materialisms and Indigenous theories of non-human agency: Making the case for respectful anti-colonial engagement. *Qualitative Inquiry*, 26(3–4): 331–346.

Smith, L.T., Tuck, E. and Yang, K.W. (eds.) (2018) *Indigenous and decolonizing studies in education: Mapping the long view*. New York: Routledge.

Steinberg, J. (2019) Spinoza's political philosophy. In E. Zalta (ed.) *The Stanford encyclopedia of philosophy*, Summer 2019 Edition. Available at: https://plato.stanford.edu/archives/sum2019/entries/spinoza-political/ [Accessed 6.7.2020].

Timmermann Korsgaard, M. (2016) An Arendtian perspective on inclusive education: Towards a reimagined vocabulary. *International Journal of Inclusive Education*, 20(9): 934–945.

Todd, Z. (2016) An Indigenous feminist's take on the ontological turn: 'Ontology' is just another word for colonialism. *Journal of Historical Sociology*, 29(1): 4–22.

Topolski, A. (2015) *Arendt, Levinas and a politics of relationality* (Reframing the boundaries: Thinking the political). London: Rowman & Littlefield International.

Voice, P. (2014) Labour, work and action. In P. Hayden (ed.) *Hannah Arendt: Key concepts*. Durham, NC: Acumen, pp. 36–51.

Watts, V. (2013) Indigenous place-thought and agency amongst humans and non humans (First woman and sky woman go on a European world tour!). *Decolonization: Indigeneity, Education and Society*, 2: 20–34.

Williams, L. (2018) *The problem with posthumansim*. Public lecture, 17th October. Research Centre for the Environmental Humanities public lecture and seminar series, Bath Spa University, UK.

Wohlleben, P. (2016) *The hidden life of trees. What they feel, how they communicate: Discoveries from a secret world*. London: William Collins.

5 Disruptive, inaugurative 'spaces of appearance'

Holding open the future

Introduction

This chapter explores ways which can *encourage*, but never guarantee, spaces of appearance in and through which new subjectivities can emerge. Such subjectivities have the potential to disrupt existing dominant Western static and separate understandings of the world and rational autonomous framings of the subject and contribute to inaugurating new ways of knowing, being and acting. I extensively draw on Western thinkers, albeit these are thinkers very much engaged with challenging such Western ideas from within.[1] These include Mouffe (2000, 2005, 2007) and her concept of agonistic pluralism, Rancière (2010a, 2010b, 2011) and his conceptions of *dissensus* and also the problem of stultification in education and Masschelein and Simons (2013) and their exploration of ways to encourage education as *Skholé*.

The chapter draws further on the ideas of Arendt writing from the borders of Western (Hellenic) and Hebraic thinking. I also draw on Andreotti, whose thinking and research is positioned at 'the interface of different knowledge systems', focusing on race, inequalities and global change and examining 'historical and systemic patterns of reproduction of inequalities and how these limit or enable possibilities for collective existence and global change' (Andreotti 2020: online). Reading these various authors together could be considered controversial, especially since both Mouffe and Rancière have openly criticised Arendt's work. However, drawing on Dikeç (2013: 1), I argue such an endeavour is worthwhile since it brings together ways to think about the possibility of education which is both 'ruptural and inaugurative'. I make links to Osberg's (2019) conception of 'symbiotic anticipation' and the possibility of radical futures as yet unimagined and unimaginable. I recognise that encouraging conditions for emergence of the new raises important ethical concerns and therefore indicate that ethics is the focus of Chapter 6.

I acknowledge that 'disruption' of one's ways of knowing, being and acting is not without discomfort and a sense of loss of 'old' certainties and ideas (see Andreotti 2016b). However, working through such feelings is a necessary step in an inaugurative process towards being together in the world in less 'harmful' ways. Support for this process is needed and

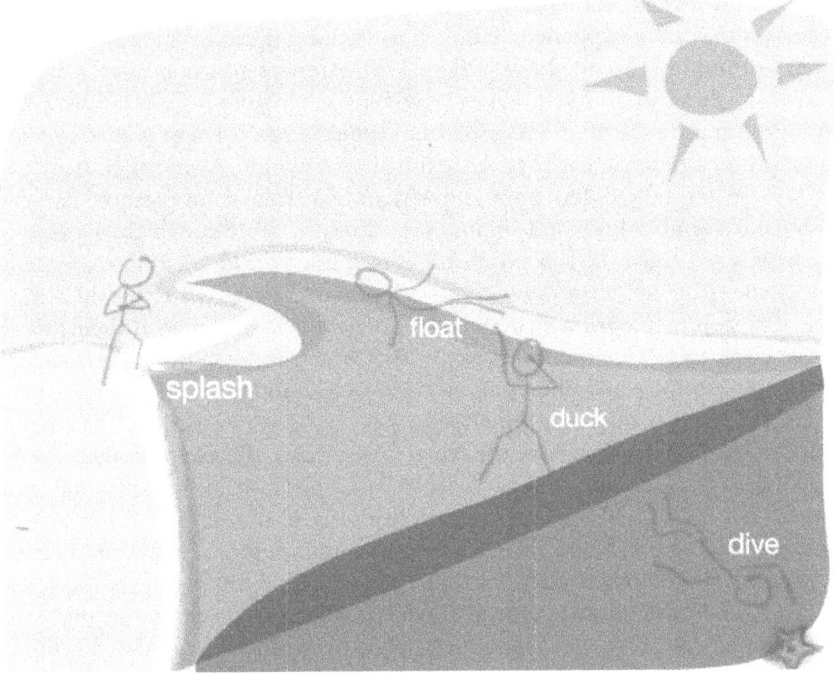

Figure 5.1 A/r/t/cart/ography #3: *The Beach*

Gesturing Towards Decolonial Futures Collective

Source: [Creative Commons]

educational settings can be a place to provide this support. The *Gesturing toward Decolonial Collective* (online) have developed the A/r/t/cart/ography *The Beach* (see Figure 5.1) 'as a response to the need to talk about our fears of "drowning" as we face the complexities, complicities, paradoxes, uncertainties, inequalities and contradictions of social and global justice work'. In this chapter I seek to encourage ways to move on from a position of 'drowning' to one of 'diving' into impossibilities, facing complexities, opening up creative potential.

Arendtian 'visiting' and opening 'spaces of appearance'

In Chapter 4 I introduced how, for Arendt (1974 [1958]: 199), 'spaces of appearance' can open up through 'acting and speaking with others who are themselves unique beings'. Such 'acting and speaking' recognises and responds to uniqueness rather than overwhelming and erasing it with preconceived ideas of 'who' the other is. The opening of spaces of appearance and the *natality* that can happen in and through such spaces have the potential to hold the future radically open since 'something new is started which cannot be expected from whatever may have happened before. This startling unexpectedness is inherent in all beginnings' (Arendt 1974 [1958]: 177). Arendt's conception of 'visiting', which she explores in *Understanding and Politics* (1994 [1954]), provides a way to approach the issue of opening spaces of appearance. 'Visiting' is an existential process (i.e. a process that happens/exists in the world). As Biesta (2013: 116) points out 'Visiting' differs from empathy in that it does not presume 'we can easily and comfortably stand in another's shoes or see through their eyes – a form of assimilation of the other into the self'. Instead, visiting retains and values the space between the self and the other. Visiting is a two-fold action. The first step is to **open a space** which allows 'us to see things in their proper perspective, to be strong enough to put that which is too close at a certain distance, so we can see and understand it without bias and prejudice' (Arendt 1994 [1954]: 323). It is this *space* which allows in, rather than silences, the unique voice of the other and enables one '*not* to see through the eyes of someone else' but instead 'to see with your own eyes from a position that is not your own' (Biesta 2013: 116). The second step is 'bridging of the abysses of remoteness until we can see and understand everything that is too far away from us as though it were our own affair' (Arendt 1994 [1954]: 323).

For Arendt it is '*imagination* alone' which enables us to engage in this two-fold action. The first step, opening a space between the self and the other, requires using one's imagination to set aside preconceived ideas and instead consider how things could be otherwise for another. The second step, 'bridging the abyss' that one has opened, requires the imagination needed to engage with, reflect on and then *understand* the lived experiences of others. Such understanding 'makes it bearable for us to live with other people in the world and makes it possible for them to bear with us' (322) – an emphasis on 'bearing with' strangers rather than trying to eradicate uniqueness.

Arendt has a particular conception of '*understanding*'. For Arendt, understanding is not the same as having correct information or scientific knowledge. Rather, understanding:

> is a complicated process which never produces unequivocal results. It is an unending activity by which, in constant change and variation, we come to terms with and reconcile ourselves to reality, that is, to try to be at home in the world.
>
> (Arendt 1994 [1954]: 307–308)

For Arendt understanding arises 'out of our lived experience of the world with others' (Hansen 2004: 4). Understanding is *immanent*. Since 'understanding is neither the unmediated expression of direct experience nor mere knowledge' (Hansen 2004: 5) it requires reflective *judgement* to 'reconcile ourselves to reality' and 'make sense of the world we share'. Arendt has an interesting perspective on '*judgement*' with an important role for emergent power. Young-Bruehl (2006: 166) comments that in Arendt's thinking:

> just as acting together 'gives the enlargement of power which none can have alone', *Urteilskraft* (which literally means judgement-power not just judgement-craft or the art of judgement) gives people an experience of the world and of other people that makes them mentally powerful.

So, for Arendt, the *power* to make judgements does not arise in contemplation or through reference to abstract notions. Rather, it emerges through engaging in encounters with unique others in a range of situations – a theme I return to in Chapters 7 and 8. These ideas resonate with the immanent ethics central to the feminist ethics of care explored in Chapter 3. For Arendt, and these feminist ethical thinkers, judgement which is *not* developed through such an immanent process will be both unrepresentative and unpersuasive to others. Moreover, it has the potential to be a form of violence.

Finally, following Arendt's line of thinking, 'visiting' unique others has the potential to enable individuals to use their 'imagination', 'understanding' and 'judgement-power' to move between standpoints and develop the *enlarged mentality* needed to develop emergent *general understanding*. In this framing, general understanding is:

> not the generality of the concept – for example, the concept 'house', under which one can then subsume various kinds of individual buildings. It is, on the contrary, closely connected with the particular conditions of the standpoints one has to go through in order to arrive at one's own 'general standpoint'.
>
> (Arendt 1982 [1970]: 43–44)

Visiting, with its emphasis on opening spaces between individuals without erasing uniqueness, has the potential to encourage 'spaces of appearance' and 'something new . . . which cannot be expected from whatever may have happened before' (Arendt 1974 [1958]: 177). It requires the development of Arendtian conceptions of imagination, understanding and judgement through immanent processes rather than through references to universal principles. This then raises the question of how 'visiting' can be encouraged in educational settings and what pitfalls and problems exist – issues I now explore.

'Visiting' in educational settings

Whilst visiting in educational settings has the potential to encourage the opening of spaces of appearance, creating conditions for visiting can be problematic. Engaging with the other can be a form of tourism – reflecting on another's situation without really 'leaving the comforts of home' (Biesta 2013: 116). This contributes to a situation in which one's own comforts and advantages *vis à vis* the other are left unchallenged. Encouraging empathy is also popular but problematic in educational settings. It can be a form of assimilation of the other into the self since it assumes one can easily put oneself in to another's shoes, unlike visiting which maintains a space between the self and the other – a 'bearing with' strangers rather than an erasure of difference.

Andreotti (2016a, 2016b) argues that 'tourism' is a prevalent construction in many Western citizenship and global educational activities, as well as in Western education more generally. 'Tourism' can refer both to positionings/mindsets in the classroom as well as actual tourism where those from the West travel in order to offer to 'help' those in 'developing countries'. Andreotti argues that unless issues are addressed critically, examining the underlying assumptions and historical and present-day forces created in these positionings, then such learning is exploitative. Highlighting Western desire to 'want change to happen on terms that do not jeopardize our perceived entitlements, securities and self-images' (Andreotti 2016a: online), she argues for encouraging what she calls *counter-intuitive learning spaces* 'that require us to move beyond the desire for self-affirmation and engage in difficult, complex, and agonistic conversations'. Like Arendt, Andreotti draws attention to the role of *imagination* – an imagination which rediscovers 'our capacity to imagine beyond boxes, fences, posturing, certainties, and safety blankets' and 'requires different questions and different vocabularies anchored in the uncertainty and precariousness of our entangled collective vulnerabilities' (2016a: online). Andreotti (2012, 2015, 2016b) has developed her HEADS UP tool to assist with this process in international service experiential learning (ISEL) and global and citizenship education (GCE) (see Table 5.1). I propose Andreotti's tool can also be adapted for a variety of educational settings to encourage the process of 'visiting' and the opening of spaces in and through which new ways to be and be together can emerge.

This is a challenging model which requires hard self-questioning. This can be difficult to do since, as previously noted, existing attitudes are often so normalised in Western cultural practices that they become 'invisible'. I therefore include two examples which highlight such 'hidden attitudes'.

The first example draws on research from Corrado (2018) who highlights the damage done by prejudiced negative global discourses on the experiences of children living across the African continent as unhappy/tragic. Citing research by the African Development Bank and by Benoite, Corrado points out that such negative discourses deny the 'majority of happy and fulfilled childhoods in the same continent' (1). This negative focus suppresses the possibility for these children to engage in the reality of their world and contribute positively to global discourses. Corrado (2018: 1) draws on a study in Kenya to argue for starting this process of emancipation in the classroom through the use of dialogical approaches which would 'equip these students with skills and platforms to voice their authenticity and exercise their creativity'. The HEADS UP, model, and the questions it provides (which could, for example, be written onto cards for use in small group work) can be used by teachers to explore with

Table 5.1 HEADS UP patterns and questions in international experiential service learning (IESL) and global and citizenship education (GCE)

Historical pattern of engagement and representation	Whose idea of development / education / the way forward?	Whose template for knowledge production in IESL and GCE?
H Hegemony (justifying dominance and supporting domination)	What assumptions and imaginaries inform the ideal of development and education in this IESL/GCE initiative?	Whose knowledge is perceived to have universal value? How come? How can this imbalance be addressed?
E Ethnocentrism (projecting the views of one group as universal)	What is being projected as ideal, normal, good, moral, natural or desirable? Where do these assumptions come from?	How is dissent addressed? How are dissenting groups framed and engaged with?
A Ahistoricism (forgetting historical legacies and complicities)	How can this imbalance be addressed? How is history, and its ongoing effects on social / political / economic relations, addressed (or not) in the formulation of problems and solutions?	How is the historical connection between dispensers and receivers of knowledge framed and addressed?
D Depoliticisation (disregarding power inequalities and ideological roots of analyses and proposals)	What analysis of power relations has been performed? Are power imbalances recognised and, if so, how are they either critiqued or rationalised? How are they addressed?	Do educators and students recognise themselves as culturally situated, ideologically motivated and potentially incapable of grasping important alternative views?
S Self-congratulatory and **Self-serving** attitude (oriented towards self affirmation/cv building)	How are marginalised peoples represented? How are those students who intervene represented?	Is the epistemological and ontological violence of certain individuals being deemed dispensers of education, rights and help acknowledged as part of the problem?
U Un-complicated solutions (ignoring the complexity of epistemological, ontological and metaphysical dominance)	Has the urge to 'make a difference' weighted more in decisions than critical systemic thinking about origins and implications of solutions?	Are simplistic analyses offered and answered in ways that do not invite people to engage with complexity or recognise complicity in systemic harm?
P Paternalism (seeking affirmation of superiority through the provision of help).	How are those at the receiving end of IESL or GCE efforts to 'make a difference' expected to respond to the 'help' they receive?	Does this IESL/GCE initiative promote the symmetry of less powerful groups and recognise these groups' legitimate right to disagree with the formulation of problems and solutions proposed?

Source: (Andreotti 2016b)

students in both African and Western educational settings why these negative portrayals of African childhood occur, how they can close down the reality of the lives of others as well as why this matters.

The second example is drawn from Martin (2012), who highlights the negative impact of viewing the other, and their situation, from a singular 'needy' viewpoint. Martin cites the example provided by a colleague, Helen, who saw a flyer asking for contributions of used school uniforms to send to the Republic of The Gambia. The flyer highlighted the poverty and poor clothing of the children there and how this is one of the many barriers they face to attending school. Potential contributors were invited to empathise with the plight of these children and drop off second-hand clothing: framing the issue from a single (Western) viewpoint of poverty and neediness. Helen comments that at one time she might have unquestioningly responded to the appeal. However, involvement in critical global education had changed her perception to one which questioned causes and definitions of poverty, the suitability of western clothing etc. Such an example could be shared in education and analysed using the HEADS UP model provided by Andreotti to encourage critical questioning of one's starting point. Visiting can open alternative spaces and other possible ways to engage with and respond to the situation: spaces demanding different imaginaries, incorporating different historical perspectives and raising pertinent epistemological and social justice questions.

There is an important distinction to be made between Arendtian 'visiting' outlined here and the thinking of Dewey. As Biesta (2004) points out, Dewey builds on education as action which leads the student and the teacher to build, through social interaction, a common understanding, a certain agreement of worldviews. However, a line of thought arising from an Arendtian conception of 'visiting' recognises that 'social interaction – in so far as it is social and not a mechanism – is not based upon agreement, identity, and consensus, but exists by virtue of difference, singularity and *dissensus*'[2] (Biesta 2004: 21). The challenge is to find ways to live with and through these differences. Whilst Arendtian visiting can never guarantee the opening of spaces of appearance it offers the hope of such spaces – spaces for living with and learning from difference and the possibilities that this can open for new shared futures. Two thinkers who can also help us explore encouraging and engaging in such spaces are Mouffe and Rancière, and it is to their ideas I now turn.

Mouffe and the issue of consensus in Arendtian thinking on visiting and opening spaces of appearance

Whilst some thinkers, such as Biesta (2004), emphasise difference, singularity and *dissensus* in Arendt's thinking on social interaction, Mouffe interprets Arendtian visiting as still having too much emphasis on consensus. For Mouffe (2007: 4), Arendtian visiting 'consists in developing the ability to see things from a multiplicity of perspectives . . . as a procedure to obtain intersubjective agreement in the public sphere through an exchange of voices, opinions and persuasion'. Mouffe (2000, 2005, 2007) argues that this is an attempt to create agonism without antagonism – a position which, for her, is not possible. She comments:

> properly political questions always involve decisions which require a choice between alternatives that are undecidable from a strictly rational point of view. The rational decision[3] made, the consensus reached, occurs in a "moment of undecidability"'.

(Mouffe 2007: online)

In such 'moments of undecidability' (and here she is drawing on Derrida – see Mouffe 2005), the decision could go in a number of directions, each of which could reasonably be argued for from the perspective of the holder of that view. It is hegemonic processes (power relations) which make one way seem the rational or consensual decision rather than just one possible way amongst many. Mouffe argues for the importance of acknowledging possible alternative ways forward which have been discarded, and of recognising that 'traces' of the discarded ideas remain in the way forward, which has been decided. It is the recognition of this situation which allows in difference – what Mouffe (2000, 2005, 2007, 2014) terms '*agonistic pluralism*' – a pluralism which does not erase antagonism but instead gives it a place. Agonistic pluralism allows for the recognition that there will always be those whose views are different from one's own; whose preferred course of action was as valid as one's own but not chosen and that such a situation is an inevitable part of the human condition. This recognition opens the possibility that those holding different views from one's own can be adversaries to be defeated but respected rather than enemies to be destroyed, silenced, converted or treated as though they did not exist in the first place. This allows such adversaries into the 'space of appearance' and encourages recognition of the hegemony at play when adversaries and their views are excluded. Andreotti's (2012, 2015, 2016b) HEADS UP model could be a starting point for identifying and challenging such hegemonic processes.

Biesta and Mouffe identify different emphases in Arendt's thinking on the degree to which 'visiting' builds towards consensual thinking or towards allowing in, living with and responding to difference. Education can be a place to explore these differing viewpoints and the issues that are foregrounded in their arguments (i.e. the issue of hegemony/power relations and the implications of trying to exclude certain viewpoints) rather than trying to see whose 'view' on Arendt is correct.

Dikeç (2013) raises another issue in relation to Mouffe's thinking. He argues that whilst Mouffe's thinking can be considered 'ruptural' of a reliance on consensus and/or a single understanding of 'the rational way forward' the question remains as to what Mouffe proposes can be done from the starting point of this disruption. This is where the ideas of Jacques Rancière (2001, 2010a, 2011) can be brought into the discussion, and, in particular, what this can mean in educational contexts.

Rancière, *dissensus* and holding open the future

In some ways, Rancière and Arendt are very different thinkers – and these differences are explored here. However, as Dikeç (2013: 83) points out, for both of them 'their politics emphasises the construction of space – for speaking and acting with others for Arendt and for setting a stage for the manifestation of *dissensus* for Rancière'. For Rancière (2010a: 69) *dissensus* is 'not a conflict of interests, opinion, or value; it is a division inserted in "common sense": a dispute over what is given and about the frame within which we see something as given'.

Rancière (2010a: 69) argues that a political subject has 'a capacity for staging scenes of *dissensus*' and for opening up the 'space of a test of verification' in the existing structuring of the sensible world (i.e. the world one senses). For him, 'the generic name for subjects that stage such tests of verification is the *demos* or people' (69). *Dissensus* is not only about disruption, it is also about inserting something new into the world, a surplus, a 'more than existed before', a rearrangement of the sensible, in which those who are normally disqualified from exercising power, do exercise it nonetheless. Thus, *dissensus* allows for the possibility of enlarging 'the space of the possible' (Davis *et al.* 2004: 4) rather than only replicating the existing arrangement of the sensible world.

In *Who is the subject of the rights of man?*, Rancière speaks about Olympe de Gouges – a woman guillotined in the French Revolution – as an example of *dissensus*. De Gouges made a declaration on the scaffold that women were excluded from voting in the assembly since they were deemed to belong in the private domestic realm and yet were at the same time 'entitled' to go to the scaffold. This was a moment where the 'bare life' of women, limited to the domestic, proved also to be a 'political' life. For Rancière, her claim was a two-fold insertion of a *dissensus* – the act of insertion itself (the guillotining of a woman) and also that a woman had her voice heard publicly. It is this two-fold manifestation of *dissensus*, and 'rearrangement of the sensible' that Rancière calls 'the political'. It interrupts what Rancière calls the 'police' or existing ordering of the world. It is important to note here that Rancière accepts that there are better and worse police orders; it is not a simple categorisation of good and bad.

Rancière (2011: 3) is critical of Arendtian categorisation of what is political, stating: 'There is no pure politics. I wrote the *Ten Theses on Politics* primarily as a critique of the Arendtian idea of a specific political sphere and a political way of life'. He argues that this categorisation divides those destined to take part in a political life located in a political sphere and participate in the Aristotelian conception of a good life from those destined merely to exist/labour or work and thus only experience 'a bare life'. Rancière argues that it is through the process of *dissensus* that those who have no place, no voice to be heard in the public domain can claim a voice rather than merely producing 'noise', affirming their 'equality as an axiom, as an assumption, and not as a goal' (Rancière *et al.* 2000: 3). The political is this polemical, disputatious process. Rancière's use of the phrase 'pure politics' in his criticism of Arendt is also important. He criticises Arendt for excluding issues relating to the labour needed for bodily functioning and issues around work (the two together make up what Arendt calls 'the social'). For Rancière (2011: 4) 'political action consists in showing as political what is viewed as "social", "economic" or "domestic". It consists of blurring the boundaries and reconfiguring who can speak'. In *Proletarian Nights* (2012), Rancière provides an example of this thinking. This text explores the life of workers who, in the 'police order', had no place to be heard as intellectuals. Yet these workers organised themselves to study at night and be productive in the arts rather than sleep. They claimed a time they had no right to and acted in ways not configured in the existing order. They inserted their intellectual work into the world, rearranging who could speak, who could have a voice as an intellectual and what counts as a political question. They also brought into question and changed what counted as intellectualism (and here one can reflect too on recognition of Indigenous knowledges as *knowledge* rather than folklore or tradition) since the workers' version of intellectualism was different from that which previously existed. For Rancière, this process of *dissensus is* democracy. As Biesta (2013: 35) comments:

> Democracy – or to be more precise, the appearance of democracy – is therefore not simply the situation in which a group that has previously been excluded from the realm of politics steps forward to claim its place under the sun. It is at the very same time the creation of a group as a group with a particular identity which did not exist before.

Rancière's criticism of Arendt's emphasis on categorisation with a special category for political action does make it seem difficult to propose drawing the ideas of Arendt and Rancière into 'conversation'. However, this difficulty requires further attention. Schaap (2011) points out that Rancière was writing from the standpoint of French intellectuals reacting against a 'version' of Arendt's ideas developed by proponents of the 1980s New French Thought movement such as Alain Renaut and Luc Ferry. These thinkers 'wished to emphasise a state-centric consensus politics' (37) and 'used Arendt to support their position' (37). I contend that this appropriation is a misuse of Arendt's ideas. As cited in Chapter 4, Arendt (1974 [1958]: 198) proposes that the public space of politics (the *polis*) is not a separate civic category; rather it 'draws on the conviction that arises out of acting and speaking together, and its true space lies between people living together for this purpose, no matter where they happen to be'.

In contrast, Honig (1995: 146), writing in an American line of thinking, emphasises the agonistic possibilities in Arendtian thinking on the public space of politics. This is *not* as a specific place, such as the Greek *topos*, but 'as a metaphor for a variety of agonistic spaces both topological and conceptual that might occasion action'. In such an understanding, Honig (1995: 146) argues:

> We might be left with a notion of action as an event, an agonistic disruption of the ordinary sequence of things that makes way for novelty and distinction, a site of resistance of the irresistible, a challenge to the normalising rules that seek to constitute, govern and control various behaviours.

In emphasising the inaugurative aspects made possible by natality rather than emphasising categories and 'pure politics', Honig provides a way to bring together Arendtian action and Rancière's notion of *dissensus*. Both share a commitment to *politics as action* and to avoiding 'an understanding of politics around given identities' (Dikeç 2013: 78). Whilst recognising that there are extensive discussions around the similarities and differences between Arendt and Rancière (for example see Schaap 2011, Beltrán 2009, Honig 1995), what has been explored here is how their ideas contribute to conditions which make 'spaces of appearance' and the holding open of the future in education a possibility. Visiting and *dissensus* both offer possibilities of holding open democratic spaces, including in education, where there is freedom to:

> call something into being which did not exist before, which was not given, not even an object of cognition or imagination, and which therefore, strictly speaking, could not be known. Action, to be free, must be free from motive on one side, from its intended goal as a predictable effect on the other.
>
> (Arendt 2006b [1961]: 150)

The Youth Climate Strike movement, however, shows the complexity of *dissensus*, particularly when it is brought into conversation with the ideas of Arendt and also those of Masschelein and Simons (2013) on education as *skholé* and the importance of protection from politicisation introduced in Chapter 4. The movement evolved from the action of Greta Thunberg. In August 2018, at the age of 15, she began striking from school every Friday in order to sit outside the Swedish parliament with a placard reading *Skolstrejk för Klimatet* (School Strike for the Climate) (Belam 2019). This was to protest about the lack of government action in responding to IPCC's[4] recommendations, based on currently known science, to reduce global warming to 1.5 degrees necessary to avoid catastrophic ecological breakdown. Her lonely protest captured the attention of young people all over the world in ways which were previously 'not even an object of cognition or imagination' (Arendt 2006b [1961]: 150) and has led to regular Friday school strike actions around the world, such as a monthly youth school strike action in Bath, UK (see Figure 5.2).

Figure 5.2 Youth Climate Strike, Bath UK Friday 22 February 2019
Source: [Author's own photo]

Adults were invited to join the worldwide Friday strike on 20 September 2019 ahead of the United Nations Climate Action Summit in New York. It is estimated that 4 million adults and young people took part in the strike that day (Belam 2019) and from 20 to 27 September 7.6 million people took part in 6135 actions in 185 countries (Global Climate Strike 2019).

This movement by children and young people can be understood as an assertion of *dissensus*: an insertion of their 'voice' into the public sphere where children have no right to speak in the existing arrangement of the sensible world. It has created something new in the world, emerging in unanticipated ways and changing the existing arrangement of the world and who has a 'voice' in it. However, there is a tension here with the conception of a radically democratic move opened by education as *skholé* (see Masschelein 2011, Masschelein and Simons 2013). As introduced in Chapter 4 the *skholé* provides free time, a space away from 'solving' the pressing issues of the world, for which adults, as Arendt (2006a: 193) argues, should be taking responsibility. Education as *skholé* can open spaces for *natality*, for as yet unknown scientific approaches and also ways of being: a democratic move towards opportunities to 'undertake something new, something unforeseen by us' (Arendt 2006a: 193). Greta Thunberg, I feel, conveyed a sense of this in her address to the United Nations Climate Summit in New York on 23 September 2019 when she declared:

> This is all wrong. I shouldn't be up here. I should be back in school on the other side of the ocean. Yet you all come to us young people for hope. How dare you! You have stolen my dreams and my childhood with your empty words.
>
> (Chappell 2019)

Tension arising from bringing ideas of these different thinkers into conversation is not necessarily a weakness. Instead of focusing on which is the 'right' approach, one can encourage a process of recognising and exploring the tensions, the overlaps and contradictions created and the potential this can generate for 'enlarging the space of the possible' (Davis *et al.* 2004: 4) and for inserting new ways to be together in the world.

Rancière, when writing specifically on education, highlights two further interconnected aspects in his thinking which, I argue, can contribute to opening spaces of appearance and to holding open the possibility of radically new futures. These are *equality as axiomatic* and *education as stultification*.

Equality as axiomatic takes the **presumption** *of equality of intelligence as its starting point*. Equality of intelligence is **not** a truth claim. Rather it is about seeing 'what can be done under that assumption' (Rancière 1991: 46). In *The Ignorant Schoolmaster: Five lessons in Intellectual Emancipation* (1991), Rancière examines the activities of Joseph Jacotot who, in 1818, found himself in exile and teaching at the University of Louvain where a number of his students spoke only Flemish and he only spoke French. Using a recently published bi-lingual edition of *Télémaque*, Jacotot asked the Flemish students, through an interpreter, to learn and repeat the French text, with the help of the Flemish translation (Flemish being their first language) then write in French (their new language) what they had read. The experiment exceeded Jacotot's expectations, raising the question: 'How could these young people, deprived of explanation, understand and resolve the difficulties of a language entirely new to them?' (2). For Rancière, this experience challenged 'what is blindly taken for granted in any system of teaching: the necessity of explication' (2), raising the question 'Why is it necessary that 'the words of the master must shatter the silence of the taught material?' (4). Moving away from a position in which explication is at the heart of teaching opens possibilities of what Rancière calls an emancipatory education in which the role of the teacher can be to 'challenge, to oblige another intelligence to exercise itself' (Rancière 2010b: 29) and to verify 'that the work of intelligence is done with attention' (29), founded on a belief in the capacity of the learner *to exercise their intelligence* in these ways. However, as Bingham and Biesta (2010: 42–43) point out, the verification envisioned by Rancière 'is not like the verification in the Socratic Method where the teacher checks that the student has arrived at a predetermined point or known knowledge'. This distinction is significant since whist the Socratic method leads to that already known to the teacher 'it is in no way a path to emancipation' (Rancière 1991: 29).

For Rancière, such emancipatory education opens new possibilities for the 'manifestation of an intelligence that wasn't aware of itself' (29). Going beyond what the teacher knows or thinks has the potential to open spaces for the appearance of new as yet unimagined subjectivities and future ways of being together. Such emancipation cannot be delivered to students: rather it is 'something which has to be claimed by students over and over again' under the conditions of 'the presumption of the equality of intelligence' (Bingham and Biesta 2010: 43). It is the *presumption of intelligence*, and what can be achieved under such conditions, which makes Rancière's notion of education distinctive from other forms including activity-based or participatory learning. The presumption of equality and what can be achieved under this assertion is a *dissensus*, a disruption of existing arrangements, and thus a *democratic move*, allowing for the possibility of education as a democratic process, education *as* democracy rather than an education *about* democracy or education practising *for* a democratic role in the future.

The emancipatory education envisioned by Rancière contrasts with education premised on 'equality as a destination', where explanation offers itself as a means to reduce the inequality: an approach which, for Rancière, introduces education as a process of *stultification* and denies 'the call for moreness' (Huebner 1999: 403): the call to move beyond present ways known by the teacher. Through stultification the student learns that his intelligence is unequal. If he does not understand he can express this, and the teacher will explain again. Thus 'the child acquires a new intelligence, that of the master's explication. Later he can be an explicator in turn. He possesses the equipment' (Rancière 1991: 8).

The process of stultification, begun in school, is also taken forward into adulthood: one continues to need and expect to have things explained and done by others. This is problematic since it can be a first step in a process of developing what Spinoza (as discussed in Chapter 4) calls *potentas* – a sense of power held by others rather than *potentia* – feeling that one can generate the power to act as a unique being, to introduce the new, to change one's environment/society.

One can reflect on whether Rancière's conception of the political as *dissensus* and Arendt's understanding of the political as opening spaces for speech, action and natality are ontologically different or whether they are much closer than that. However, *in practice*, reading the two thinkers together is valuable. Understanding students as 'beginners and equals', to use Dikeç's phrase, and practicing education under this presumption, is a powerful starting point for teachers and students who wish to encourage conditions which can help open *'spaces of appearance'* and hold open the future as a 'site of radical novelty' (Facer 2016: 58).

In education today, including education exploring sustainability, there is often an endless busyness, and no space to pause and reflect. Space is an important element of 'spaces of appearance', and one that should not be overlooked. Space has the potential to be powerful and profound. Reflecting on his experiences of the Holocaust, Frankl (2004 [1957]) identified 'space' as crucial to his survival. He realised that, like him, many of his fellow survivors were not the physically fittest. Rather they were those who were able to create a mental space in which they could reflect and choose how they responded to the horrors around them, to develop even in the most desperate and seemingly powerless situations a sense of *potentia* rather than *potentas*.

As introduced in Chapter 4, and earlier in this chapter, Masschelein and Simons (2013) also argue for the importance of space in education as *skholé* – education which opens space to pay attention in the present, space to question and challenge the existing ordering of things, space for encounters through which one's uniqueness and one's *potentia* to bring in the new and unexpected can emerge. In their thinking they also provide ideas to *encourage* such education as *skholé*, as the next section explores.

Masschelein and Simons – encouraging possibilities for education as *skholé* and how this can contribute to 'spaces of appearance' in education

Masschelein and Simons (2013) propose the steps of *suspension*, *profanation* and *attention* as ways to encourage education as *skholé*.

Suspension is a matter of suspending, at least for a short while, the child's past, family problems, terminal illness and *responsibility* to solve today's worldly problems. It is a temporary suspension of these constraints for the student, the teacher and the subject matter in order that school can be a place of transformation. To reinforce here the point made in Chapter 4, this is not encouraging a space in which children are 'protected' from exploring challenging worldly issues of the past and present. Rather it is about protecting children from being used as political tools and also protecting a space for them to explore such issues without the responsibility of solving them. As cited in Chapter 4,

> This does not mean dropping all prior values and throwing away the lessons of the past. It means, rather, using the lessons of the past to invent something radically new; something which might accompany us into the future (and also which might not).
>
> (Osberg 2010: 164)

Suspension opens a 'breach in linear time' (36), interrupting past and future. Drawing on Pennac (2010), Masschelein and Simons (2013: 34) emphasise that a teacher 'working the room' can draw students into the 'here and now', allowing students to 'detach from the past (which weighs them down and defines them in terms of their [lack] of ability/talents)'. (I recognise there are some problematic aspects to these ideas, as I go on to explore later.) Suspension also enables students to detach from any predetermined futures. It encourages a space which is not [only] concerned with linear time, expectations and duties and cause-and-effect thinking – a sense that 'if you do this, you will achieve that'. It opens possibilities in all directions. This does, however, require a letting go of the child into the unsettling in-between space of education – a space inserted between home and the world – if children and young people are to find their own futures. This poem by C. Day Lewis (1967]) captures this very well.

Walking Away in *Selected Poems* (1967) [Reproduced with permission]:

> It is eighteen years ago, almost to the day –
> A sunny day with the leaves just turning,
> The touchlines new ruled – since I watched you play
> Your first game of football then, like a satellite
> Wrenched from its orbit, go drifting away.
>
> Behind a scatter of boys, I can see
> You walking away from me towards the school
> With the pathos of a half-fledged thing set free
> Into a wilderness, the gait of one
> Who finds no path where the path should be.
>
> That hesitant figure, eddying away
> Like a winged seed loosened from a parent stem
> Has something I never quite grasp to convey
> About nature's give and take – the small, the scorching
> Ordeals which fire one into irresolute clay.
>
> I have had worse partings, but none that so
> Gnaws at my mind still. Perhaps it is roughly
> Saying what God alone could perfectly show -
> How selfhood begins with a walking away,
> And love is proved in the letting go.

Before moving on from Masschelein and Simons' thinking on suspension, there is an important issue which needs to be problematised. Masschelein and Simons argue that suspending *external* factors such as the child's past and family background can encourage a space through which new possibilities can emerge. What still needs to be highlighted is that such spaces are not neutral, even if some teachers and learners might believe that they are. They are spaces framed by Western worldviews (in which Masschelein and Simons are themselves situated, as emphasised by their own drawing on ideas from Ancient Greece, the source of dominant Western philosophical thinking). Other ontological and epistemological starting points such as the Indigenous and Hebraic ways of being and knowing explored in this book have also already been 'suspended'. This suspension silences, alienates and harms students for whom these ideas are significant. They learn that often their contributions remain unacknowledged, dismissed or derided.

Bhabha's (1994, 2006) conception of 'third space' can help here. The 'third space' is not some kind of physical space. Rather it is an intersection in time and space: a mode of speaking and acting which is both productive, and not merely reflective, engendering new possibilities. In such spaces:

> What is theoretically innovative, and politically crucial, is the need to think beyond narratives of ordinary and initial subjectivities and to focus on those *moments or processes that are produced in the articulation of cultural differences* [my italics]. These 'in-between' spaces provide the terrain for elaborating strategies of selfhood – singular or communal – that initiate new signs of identity, and innovative sites of collaboration, and contestation, in the act of defining society itself.
>
> (Bhabha 1994: 2)

Here Bhabha is naming an important distinction between *cultural diversity* and *cultural difference*. For him:

> Cultural diversity is an epistemological object – culture as an object of empirical knowledge – whereas cultural difference is the process of the *enunciation* of culture as "knowledgeable," authoritative, adequate to the construction of systems of cultural identification.
>
> (2006: 157)

Cultural difference is not studying about x or y cultures as objects. Instead it is the *articulation* and *enactment* of culture practices and a recognition that these practices can authoritatively claim and construct new ways of being.

Third spaces are 'hybrid spaces'[5] where cultural meaning and representation have no 'primordial unity or fixity' (Bhabha 1994). Cultural meaning and representation emerge in and through *enunciation* and *processes* in that space. 'Suspension', if handled sensitively, with openness and awareness of the taken-for-granted framings and power differentials still present in supposedly 'neutral spaces', offers opportunities for the opening of third spaces and for the articulation of cultural difference, a move similar to what Masschelein and Simons (2013: 38) call 'a matter of profanation'.

Profanation is a step which comes after suspension. The terms 'profane' and 'profanation' are used in a non-religious sense, as 'something detached from its regular use, no longer sacred or occupied by a specific meaning, and so something in the world that is both accessible to all and subject to re-appropriation of meaning' (38). Topics can be examined in ways no longer appropriated by the older generation (in particular those in the Western academic tradition) but instead made available for the younger generation for free and novel[6] use. As Pennac (2010) highlights, in school there is always something 'on the table', be it a car part, a clothing pattern, a mathematical proof, a text. This opens possibilities to pay attention to the item under scrutiny and to challenge 'the rules of the game' usually imposed in society when engaging with that particular text or item (for example 'theatre' is like this . . . 'evening clothes' are like this, but not like that). Now *unhanded* from the item's usual/appropriate social use, it invites us to explore and engage with it and, echoing Arendt, from such starting points students can begin something new.

Attention is the third step in Masschelein and Simon's conception of *skholé*. Suspension and profanation make *attention* to the topic of study possible. The *skholé* is not a place which remains remote from the world. Instead, it is possible to 'open up to the world at school' (42). For example, a student who has a particular 'everyday' experience and knowledge of birds can suspend her usual knowledge/attitudes. By *attending* to birds as the topic under consideration, she approaches birds anew with fresh eyes and can play with the ideas that the topic presents. She can use these ideas to re-orientate herself to her existing knowledge of, and relationship with, birds in the world. In addition, if teachers and students can challenge the exclusion of non-dominant worldviews from supposedly neutral educational spaces, this opens further possibilities. Students whose starting points for understandings of/attitude towards birds (or trees, or water or fish) come from non-Western (Eurocentric) ontological and epistemological systems, such as those outlined in earlier chapters, can both attend to and share these, further opening potential for new ways of knowing and being with birds.

Masschelein and Simons (2013: 43) argue that the process of *skholé* opens possibilities for emergence of new ideas, language and practices. Although the immediate utility of these things may not be apparent, these 'somethings' begin to become part of our world and begin to form us (in the sense of the Dutch word *vorming*). Such formation is not an auxiliary activity at school, as something existing *outside* the topic or item of study. Instead, formation arises in and through attention to the world as it exists in the topic of study as well as 'attention and interest for the world', and 'attention and interest for the self in relation to the world' (Masschelein and Simons 2013: 44–45). Such moments of formation have potential to open spaces in and through which who one is as a human subject can emerge *through encounters within* the world. These can be spaces for *natality*, for acting as an *initium*, a beginner, to reference Arendt, and, to reference Rancière, spaces for *dissensus* – the insertion into the existing world of something new, which has no 'right' to be under the existing arrangement of the world and which then also changes the existing world.

Abundant possibilities

Education as *skholé*, can be understood as a 'pedagogy of the present' (Facer 2016). Such a pedagogy opens 'a distinctive temporality of its own' (59), keeping in play the abundant possibilities available in the present and opening new possibilities for the future (rather than a colonisation of the future by existing ideas). This brings to mind Osberg's 'symbiotic anticipation' – a type of anticipation which does not anticipate a particular future but instead 'encourages a fusion of mutual inspiration and experimentation which opens up the boundless, incalculable possibilities which are not yet imagined or imaginable' (2019: 1472). Such education has the potential to be a radically democratic education which does not only include those excluded before – albeit that is also important. Instead, it also allows for acts of *dissensus* and for natality: reconfigurations of how the world, and ways to be in it, are understood and enacted. As previously cited, Facer (2016: 60) emphasises:

> It is in the reframing of democratic education as a politics of disclosing and holding open new possibilities *for all future generations* [original italics] rather than a realising of the pre-defined dreams of today's generation of adults, that a different form of educational project becomes available.

In this chapter, I have explored a range of theoretical ideas and some practical examples which have potential to *encourage* the opening of spaces of appearance in educational settings. In Chapters 7 and 8, I include further theoretical and practical examples and case studies. My intention is to reflect a 'patterning' approach to the way that knowledge and meaning can emerge in 'a 'discursive exchange' between theory and practice (Reiss 1982: 30 cited in Trueit 2005: 31). However, before doing this, I recognise the need to address how encouraging opening spaces of appearance raises the important issue of ethics. What if ways which emerge are deemed unethical, and who gets to decide? In the next chapter, I draw on Arendt's conception of forgiveness and mutual promising to consider these questions and the need for emergent, immanent ethics which respond to the unboundedness, irreversibility and unexpectedness which arise when one's own natality, one's beginnings, are inserted into the world and taken up by others.

Notes

1 I recognise that there are many ways in which Indigenous thinkers can contribute to this process. I feel it is my role to encourage a direct reading of/engagement with these Indigenous thinkers (and I provide some possible starting points for this in Chapters 2 and 3) rather than attempt to paraphrase them – a paraphrasing limited by my own 'Western' ways of understanding the world. As Bonnett (2000, 2002, 2004) points out, it is hard, if not impossible, to break out of one's own metaphysical framings even if one thinks one has. Little Bear (2011, 2016) recognises this difficulty and suggests a good starting point for engaging with different metaphysical framings is through learning the language of a particular Indigenous people. He gives as an example how in his Blackfoot language there

is much more emphasis on verbs, emphasising process rather than nouns used to classify the world as in Western languages.
2 Here Biesta is using Rancière's term *dissensus*, which I explore later in the chapter.
3 Mouffe also asserts that Habermas' approach to deliberative decision making is reliant on an appeal to an independent 'rational truth'. She then draws on Wittgenstein to show how 'rational truths' are in fact situated in different understandings of the world and that a single 'rational truth' is not possible (see Mouffe 2005).
4 The Intergovernmental Panel on Climate Change (IPCC) is the United Nation's body for assessing the science relating to climate change.
5 See Meredith (1998) for further discussion of idea of hybridity in the third space.
6 'New' here can include those ideas which have existed for thousands of years but have been excluded and often denigrated by dominant cultures (i.e. 'new' includes 'new ways' to recognise these ideas); see St Pierre (2016), Todd (2016). Macfarlane (2019) calls such ideas 'new-old', making this issue more visible.

References

Andreotti, V. (2012) Editor's preface: HEADS UP. *Critical Literacy: Theories and Practices,* 6(1): 1–3. Available at: www.oregoncampuscompact.org/uploads/1/3/0/4/13042698/andreotti_-_preface_-critical_literacy_org_-_headsup__1_.pdf [Accessed 23.1.2020].

Andreotti, V. (2015) Global citizenship education otherwise: Pedagogical and theoretical insights. In A. Abdi, L. Shultz and T. Pillay (eds.) *Decolonizing global citizenship education.* Rotterdam: Sense Publishers, pp. 221–230.

Andreotti, V. (2016a) Multilayered selves: Colonialism, decolonization and counter-intuitive learning spaces. *Arts Everywhere, Musagetes.* Available at: http://artseverywhere.ca/2016/10/12/multi-layered-selves/ [Accessed 11.8.2018].

Andreotti, V. (2016b) The educational challenge of imagining the world differently. *Canadian Journal of Development Studies Revue Canadienne d'études du développement,* 37(1): 101–112.

Arendt, H. (1974 [1958]) *The human condition.* Chicago, IL: University of Chicago Press.

Arendt, H. (1982 [1970]) *Lectures on Kant's political philosophy.* Beiner, R. (ed.). Chicago, IL: University of Chicago Press.

Arendt, H. (1994 [1954]) Understanding and politics. In *Essays in understanding.* New York: Schocken Books.

Arendt, H. (2006a [1961]) *Crisis in education. In between past and future: Eight exercises in political thought.* London: Penguin.

Arendt, H. (2006b [1961]) *What is freedom? In between past and future: Eight exercises in political thought.* London: Penguin.

Belam, M. (2019) Greta Thunberg: Teenager on a global mission to 'make a difference. *The Guardian,* 26th September. Available at: www.theguardian.com/environment/2019/sep/26/greta-thunberg-teenager-on-a-global-mission-to-make-a-difference [Accessed 29.10.2019].

Beltrán, C. (2009) Going public: Hannah Arendt, immigrant action and the space of appearance. *Political Theory,* 37(5): 595–622.

Bhabha, H.K. (1994) *The location of culture.* Abingdon, Oxon: Routledge.

Bhabha, H.K. (2006) Cultural diversity and cultural differences. In B. Ashcroft, G. Griffiths and H. Tiffin (eds.) *The post-colonial studies reader.* New York: Routledge, pp. 155–157. Available at: http://monumenttotransformation.org/atlas-of-transformation/html/c/cultural-diversity/cultural-diversity-and-cultural-differences-homi-k-bhabha.html [Accessed 24.10.2019].

Biesta, G. (2004) "Mind the gap!" Communication and the educational relation. In C. Bingham and A.M. Sidorkin (eds.) *No education without relation*. New York: Peter Lang, pp. 11–22.

Biesta, G. (2013) *The beautiful risk of education*. London: Paradigm Publishers.

Bingham, C. and Biesta, G. (2010) *Jacques Rancière: Education, truth and emancipation*. London: Continuum.

Bonnett, M. (2000) Environmental concerns and the metaphysics of education. *Journal of the Philosophy of Education*, 34(4): 591–602.

Bonnett, M. (2002) Education for sustainability as a frame of mind. *Environmental Education Research*, 8(1): 9–20. doi: 10.1080/13504620120109619

Bonnett, M. (2004) *Retrieving nature: Education for a post-humanist age*. London: Wiley.

Chappell, B. (2019) 'This is all wrong,' Greta Thunberg tells world leaders at U.N. climate session. *NPR*, 23rd September. Available at: www.npr.org/2019/09/23/763389015/this-is-all-wrong-greta-thunberg-tells-world-leaders-at-u-n-climate-session [Accessed 31.10.2019].

Corrado, E. (2018) *Using dialogue – spaces to develop African children's autonomy: A Kenyan Study*. Paper presented at VIII International Conference on Critical Education, 25–28th July, London, UK.

Davis, B., Phelps, R. and Wells, K. (2004) Complicity: An introduction and a welcome. *Complicity: An International Journal of Complexity and Education*, 1(1): 1–7.

Dikeç, M. (2013) Beginners and equals: Political subjectivity in Arendt and Rancière. *Transactions of the Institute of British Geographers*. Royal Geographical Society (with the Institute of British Geographers), 38(1): 78–90.

Facer, K. (2016) Using the future in education: Creating space for openness, hope and novelty. In H.E. Lees and N. Noddings (eds.) *The Palgrave international handbook of alternative education*. London: Palgrave Macmillan, pp. 53–78.

Frankl, V. (2004 [1957]) *Man's search for meaning: The classic tribute to hope from the Holocaust*. London: Rider.

Gesturing toward Decolonial Futures Collective (online) *A/r/t/cart/ography #3: The beach*. Available at: https://decolonialfutures.net/portfolio/the-beach/ [Accessed 3.2.2019].

Global Climate Strike (2019) *Global climate strike,* 20–27th September. Available at: https://globalclimatestrike.net/ [Accessed 29.10.2019].

Hansen, P. (2004) Hannah Arendt and bearing with strangers. *Contemporary Political Theory*, 3: 3–22. doi: 10.1057/palgrave.cpt.9300124

Honig, B. (1995) Toward an agonistic feminism: Hannah Arendt and the politics of identity. In B. Honig (ed.) *Feminist interpretations of Hannah Arendt*. Philadelphia: Pennsylvania State University Press, pp. 135–166.

Huebner, D. (1999) *The lure of the transcendent*. Mahwah, NJ: Erlbaum.

Lewis, C.D. (1967) Walking away. In *Selected poems*. New York: Harper and Row.

Little Bear, L. (2011) *Native science and western science: Possibilities for collaboration*. Lecture on 4th of March at Arizona State University. Available at: https://www.youtube.com/watch?v=ycQtQZ9y3lc [Accessed 15.11.2018].

Little Bear, L. (2016) *Blackfoot metaphysics 'waiting in the wings'*. Congress of the Humanities and Social Sciences Big Thinking Lecture 1st June. Available at: https://www.youtube.com/watch?v=o_txPA8CiA4 [Accessed 18.11.2018].

Macfarlane, R. (2019) Should this tree have the same rights as you? *The Guardian*, 2nd November. Available at: www.theguardian.com/books/2019/nov/02/trees-have-rights-too-robert-macfarlane-on-the-new-laws-of-nature [Accessed 7.11.2019].

Martin, F. (2012) Thinking differently about difference. *Think Global Thinkpiece, 2012 Series*. Available at: https://files.eric.ed.gov/fulltext/ED564429.pdf [Accessed 21.2.2020].

Masschelein, J. (2011) *Experimentum scholae*: The world once more . . . but not (yet) finished. *Studies in the Philosophy of Education*, 30(5): 529–535. doi: 10.1007/s11217-011-9257-4

Masschelein, J. and Simons, M. (2013) *In defence of schools: A public issue*. Leuven: Education, Culture and Society Publishers. Available at: http://ppw.kuleuven.be/ecs/les/in-defence-of-the-school/masschelein-maarten-simons-in-defence-of-the.html [Accessed 3.2.2016].

Meredith, P. (1998) *Hybridity in the third space: Rethinking bi-cultural politics in Aotearoa/New Zealand*. Paper Presented to Te Oru Rangahau Maori Research and Development Conference, 7–9th July, Massey University. Available at: http://lianz.waikato.ac.nz/PAPERS/paul/hybridity.pdf [Accessed 24.10.2019].

Mouffe, C. (2000) *Deliberative democracy or agonistic pluralism* (Political science series). Vienna: Institute for Advanced Studies. Available at: www.ihs.ac.at/publications/pol/pw_72.pdf [Accessed 2.12.2016].

Mouffe, C. (2005) *The democratic paradox*. New York: Verso.

Mouffe, C. (2007) Artistic activism and agonistic spaces. *Art and Research: A Journal of Ideas, Contexts and Methods*, 1(2). Available at: www.artandresearch.org.uk/v1n2/mouffe.html [Accessed 9.2.2015].

Mouffe, C. (2014) Democracy, human rights and cosmopolitanism: An agonistic approach. In C. Douzinas and C. Gearty (eds.) *The meaning of rights: The philosophy and social theory of human rights*. Cambridge: Cambridge University Press, pp. 181–192.

Osberg, D. (2010) Taking care of the future. In *Complexity theory and the politics of education*. Rotterdam: Sense, pp. 153–166.

Osberg, D. (2019) Education and the future: Rethinking the role of anticipation and responsibility in multicultural and technological societies. In R. Poli (ed.) *Handbook of anticipation*. Cham, Switzerland: Springer International, pp. 1459–1478.

Pennac, D. (2010) *School blues*. Ardizzone, S. (tr.) London: MacLehose Press.

Rancière, J. (1991) *The ignorant schoolmaster: Five lessons in intellectual emancipation*. Ross, K. (tr.). Redwood City, CA: Stanford University Press.

Rancière, J. (2001) Ten theses on politics (in English). *Theory & Event*, 5(3).

Rancière, J. (2010a) Who is the subject of the rights of man? In J. Rancière (ed.) *Dissensus: On politics and aesthetics*, pp. 70–83. London: Continuum.

Rancière, J. (2010b) On ignorant schoolmasters. In C. Bingam and G. Biesta with Jaques Rancière (eds.) *Jaques Rancière: Education, truth, emancipation*. London: Continuum.

Rancière, J. (2011) The thinking of dissensus: Politics and aesthetics. In P. Bowman and R. Stamp (eds.) *Reading Rancière*. Continuum: London, pp. 1–17.

Rancière, J. (2012) *Proletarian nights*, 2nd Edition. New York: Verso.

Rancière, J., Guenoun, S., Kavanagh, J. and Lapidus, R. (2000) Jaques Rancière: Literature, politics, aesthetics – approaches to democratic disagreement. *SubStance*, 29(2): 3–24. doi: 10.2307/3685772

Reiss, T.J. (1982) *The discourse of modernity*. Ithaca, NY: Cornell University Press.

Schaap, A. (2011) Enacting the right to have rights: Jacques Rancière's critique of Hannah Arendt. *European Journal of Political Theory*, 10(1): 22–45. doi: 10.1177/1474885110386004

St Pierre, E. (2016) Curriculum for new material, new empirical inquiry. In N. Snaza, D. Sonu, S.E. Teruman and Z. Zaliwska (eds.) *Pedagogical matters: New materialisms and curriculum studies*. New York: Peter Lang, pp. 1–12.

Todd, Z. (2016) An Indigenous feminist's take on the ontological turn: 'Ontology' is just another word for colonialism. *Journal of Historical Sociology,* 29(1): 4–22.

Trueit, D.L. (2005) *Complexifying the poetic: Toward a poiesis of curriculum.* Available at: https://digitalcommons.lsu.edu/cgi/viewcontent.cgi?article=3987&context=gradschool_dissertations [Accessed 19.9.2020].

Young-Bruehl, E. (2006) *Why Arendt matters.* New Haven: Yale University Press.

6 Arendt's forgiveness and mutual promising

A framework for approaching ethical issues opened by emergence

Introduction

Central to this book is that emergence of the new is essential for sustainability as all ecosystems, with humans included as part of such systems, need to adapt and change in order both to survive and do well. This includes emergence of new ways to be a subject. Opening spaces for the emergence of the new, however, is not unproblematic. What emerges may be seen as undesirable or dangerous, further raising the question of who gets to decide what is desirable. Moreover, such a sharp binary between whether something is deemed 'good' or 'bad' is questionable – a situation, for example, which Pierre and Marie Curie had to grapple with in their scientific work around radioactivity and its subsequent side effects and uses in ways that no one could foresee. Another complicating factor is that what is heralded as 'new' could have existed in non-dominant cultures for many centuries. Naming something as 'new' thus risks being a new form of universalism and domination (for example see Todd 2016) when this 'new' way is 'rolled out' across different places. Encouraging emergence opens the '*unpredictability*', '*irreversibility*' and '*unboundedness*' which arise when new ways are inserted into the world and taken up by others. Yet, if humanity is to avoid 'colonisation' of the future (Facer 2016) with ways of being and ideas of the present, allowing the radically new to emerge is necessary, and approaches need to be found to respond to the ethical challenges that such emergence inevitably opens up.

In the *Human Condition* (1974 [1958]), Arendt develops her *two-fold concept of forgiveness and mutual promising* which provides ways to approach the risks and ethical issues which emergence generates. This is an *immanent approach to ethics* arising through people speaking and acting with others under conditions of plurality.[1] As I go on to explore, it is an approach which resonates with the relational feminist ethics of care and immanent ethics, for example of Gilligan, Held and Noddings, the web of relations emphasised by Tronto and Fisher; the ethics identified by Little Bear, De Line and Donald arising from the interconnection and intraconnection central to the Indigenous concept of *All my/our Relations* and the 'responsibilities that come with being in relation' (Donald *et al.* 2012: 71).

In this chapter, I draw on Topolski (2011) to argue that in her approach to ethics Arendt *builds* on the dynamic sense of turning from past wrong-doing inherent in the Hebraic process of *teshuvah*.[2] *Teshuvah* emphasises *shuv* – the act of turning. Arendt highlights the roles of forgiveness and mutual promising between people as *intrinsic* to such a process of turning, albeit Arendt is using these ideas in a political space emphasising plurality, and not in the religious sphere. Topolski argues that Arendt is *developing* ideas of *teshuvah* in a space which Topolski calls 'the Judaic' – an intellectually rich space between Jewish (Hebraic) religious thinking and Hellenic/Western philosophy. Arendt is writing from the richness of the 'borderlands'[3] with the potential to bring different ways of thinking and being into conversation and generate new ways of being in the world together. Topolski highlights that such a move is problematic for Western scholars who are conditioned through 'the canonised Platonic model of political philosophy' (6) to draw 'sharp boundaries between the realms of politics, ethics and faith' (6). Exploring how power to forgive and make mutual promises on specific issues in specific contexts arises horizontally between people is thus not an accidental conflation of ethics and politics. It is a deliberate argument and one which emphasises how Arendt's orientation towards the political arises from the ethical dilemma of how to live with others, how to 'bear with others, strangers forever in the same world' (1994 [1954]: 322), without falling prey to totalitarianism and all that it can, and has, unleashed.

Bernstein (2002) argues that Arendt's concern with the dangers and impact of totalitarianism is a 'thought train' which runs through and unites her work. Whilst not wanting to limit Arendt's thinking to being a response to her experiences of the rise of totalitarianism in the 1930s and to the Holocaust, Bernstein notes their significant impact on the direction her work took. Topolski (2015) and Botbol-Baum and Roviello (2013) argue that whilst both Arendt and the ethical philosopher Levinas were students of Heidegger and influenced by his ideas, the catastrophic events of the Second World War and the Holocaust caused an *arrachement* – a wrenching/tearing away or abruption – of their ideas from those of Heidegger and from 'philosophy of singularity' generally (Topolski 2015). Whilst Heidegger's focus was on 'Man', Arendt's and Levinas' concern was for *being together with others ethically* and '*what arises in the space between I and we*' (Topolski 2015: 176, emphasis added). Topolski calls this a concern for *Mittsein* rather than *Dasein*.

I acknowledge that claiming an ethical turn in Arendt's thinking is a bold move, since Arendt is usually categorised as a political thinker, with Birmingham (2006: 131 cited in Topolski 2011: 7) going so far as to state 'the most difficult aspect of Arendt's political theory is her insistence that the ethical be removed from political thinking'. In this chapter I highlight that whilst Arendt always argued strongly against a *normative*[4] ethical approach in the political sphere, she did, through her two-fold approach of forgiveness and mutual promising, argue for the importance of an *immanent* horizontal ethics emerging between people speaking and acting together.

For Arendt, the forgiveness which *teshuvah* calls for allows a turning and *release from,* but *not a reversal of,* the past and trespasses committed there. Mutual promising 'requires doing something . . . acting politically by means of words and deeds to create a shared world' (Topolski 2011: 7), strengthening 'the sometimes-fragile bonds between people' (7) and opening possibilities of hope of novelty in the future. Moreover, since forgiveness rather than vengeance is unexpected and enables us to demonstrate our *freedom* to choose how to act, the process of forgiveness and mutual promising can be understood as a *democratic move.* Forgiveness, however, is challenging, and not something Arendt underestimates, especially in instances of what Arendt terms 'extreme evil'. In this chapter, I therefore examine the issue of extreme evil, how it can develop and the possible limits to forgiveness extreme evil presents.

There is now an urgent temporal aspect to ethical engagement with emergence of the new, especially since much of what has emerged as 'new' in recent times is causing harm. Ways forward need to be found which can both deal with what has happened and find new less harmful ways to move forward. As Hans Jonas foresaw, we live in an unprecedented era creating what he calls an *imperative of responsibility* (1984) to think about our ethical relationships not only in the present but also in and with the future. In previous times humans' powerful and sometimes harmful forays into the wider world made marks upon it, but the wider natural world could, overall, absorb these: 'the earth was deathless' (*Antigone,* cited in Jonas 1984: 2). Human action played out in a world which changed but remained fundamentally 'undamaged' by human actions informed by various principles of good and evil. Enjoinders 'to do' or 'not to do' were built on a focus of the ethical impact of one's actions in the present. Foretelling the Anthropocene, Jonas argued that modern technology has now changed humanity's relationship with the wider world. What some of humanity does now is already damaging and has potential to damage further the world in profound and permanent ways. Thus, now, more than ever before, ethical frameworks are needed that can help humanity cope with the unpredictability, unboundedness and irreversibility of new ways of being, knowing and acting in the world – a framework which Arendt provides.

Throughout the chapter, I include case studies and examples drawn from a variety of situations to aid engagement with the theoretical ideas explored, provide ideas for practical discussion and development with students and also encourage new theorisation. The case studies have been selected so they can be approached in different ways, acknowledging that factors such as students' ages must be handled sensitively when considering ethical issues, and that students from different locations bring different perspectives and starting points to the ideas discussed.

Arendt's forgiveness and mutual promising – a two-fold action

Arendt develops her conception of forgiveness and mutual promising as a way to respond to the ethical issues generated by the unpredictability, unboundedness and irreversibility of the new in various texts, for example in *The Jewish Writings*[5] (2007), *The Human Condition* (1974 [1958]) and *On Revolution* (1973 [1963]). It forms an integral part of her concept of action. For Arendt (1974 [1958]: 237):

> Without being forgiven, released from the consequences of what we have done, our capacity to act would, as it were, be confined to one single deed from which we would never recover; we would remain the victim of its consequences for ever, not unlike the sorcerer's apprentice who lacked the magic formula to break the spell.

On mutual promising, Arendt (1974 [1958]: 237) comments:

> Without being bound to the fulfilment of promises, we would never be able to keep our identities; we would be condemned to wander helplessly and without direction in the darkness of each man's lonely heart, caught in its contradictions and equivocalities – a darkness which only the light shed over the public realm through the presence of others, who confirm the identity between the one who promises and the one who fulfils, can dispel.

Plurality is central to forgiveness and promising since, as Arendt explains in this quotation, one cannot forgive oneself nor be bound to a promise only made to oneself in isolation. Arendt draws a distinction between her thinking and thinking ruled by Platonic conceptions based on a relationship between 'me and myself'. To explain this, she points out that in a Platonic (Western) philosophical approach to morality, based on singularity, the starting point is the self – 'how one rules himself, he will rule others' (1974 [1958]: 238). Arendt thus also departs from Kant, whose conception of morality, building on a Platonic approach, is based on autonomy (a Western philosophical conception of subjectivity which this book calls into question). *For Arendt, morality is rooted in plurality, in action between people in the public domain.* It is a process requiring *interconnected action from both.* This idea is central in this chapter.

Commentators such as Birmingham (2006) and La Caze (2014) emphasise the political nature of Arendt's thinking on forgiveness and mutual promising explored in *The Human Condition*. However, as introduced at the start of this chapter, Arendt's writing stands *hors catégorie*, arising as it does from both Western philosophy and political theory and from Jewish religious thinking. Whilst Arendt does draw extensively on Greek and Roman political models, her project is not a Greek or Roman one, although it is often (mis)understood as such. Instead these models are starting points to 'guide Arendt's own analysis and provide her with material from which she can distil anew the authentic meaning of politics' (Jakub 2005: 2). For Arendt, forgiveness and mutual promising 'belong together' (Arendt 1974 [1958]: 237) and form into a single concept. *Teshuvah* calls for forgiveness, but a forgiveness which cannot simply be translated in the psychological sense of 'I forgive you'. In Arendt's thinking, *Teshuvah* involves a sense both of a move away from wrong-doing and a turning towards a demand-ing commitment (promise) by *all* (not just the person who has done wrong) to act in new ways, ways *which include a recognition and redressing of wrongs committed*, in the horizontal relationship engendered by forgiveness and mutual promising. It emphasises *the action of turning*, arising:

> directly out of the will to live together with others in the mode of acting and speaking together, and thus . . . they are like control mechanisms built into the very faculty to start new and unending processes.
>
> (Arendt 1974 [1958]: 246)

It is this *horizontal relationship* between people which is the source of a dynamic *horizontal ethics* which:

> arises from a plurality, and in the between, rather than coming from with-out or above. A relational ethic, rooted in a 'relative' and horizontal tran-scendence, helps us to strengthen the fragile human realm by providing a web of relations with faith and hope, both of which are fundamental to the political.
>
> (Topolski 2011: 8–9)

Thus, understanding forgiveness and mutual promising as *a two-fold action* of *shuv*, as immanent horizontal ethics arising between people which allows a *release* from past acts and a turn towards hope for novelty in the future, can make an important contribution to the problematic issue of ethics opened up by allowing space for emergence.

In order to explore Arendt's two-fold action of forgiveness and mutual promising, it is necessary to look at each aspect in more detail whilst always bearing in mind that both aspects are part of a dynamic whole which 'belong together' (Arendt 1974 [1958]).

The aspect of forgiveness

Forgiveness

Arendt outlines how the remedy against the irreversibility of new actions gen-erated by emergence lies with the potentiality of action itself. For Arendt:

> The possible redemption from the predicament of irreversibility – of being unable to undo what one has done, though one did not and could not, have known[6] what he was doing – is the faculty of forgiving . . . forgiving serves to undo the deeds of the past, whose sins hang like Damocles' sword over every new generation.
>
> (Arendt 1974 [1958]: 237)

For Arendt, forgiveness is not (only) a personal act but a key aspect of political life lived with others. The phrase 'serves to undo' could be misunderstood as a suggestion of a reversal of time – a position incompatible with the complexity and complexity-compatible framing underpinning this book. However, it is important to note here that Arendt uses the phrase 'serves to undo' not 'undoes'. For her, the redemptive force of forgiveness is not an escape from, or closure of the past through a reversal of a past act. Instead, as Arendt clarifies (1974 [1958]: 240), mutual forgiveness enables *a release*[7] from past misdeeds and:

> Only through this constant mutual release from what they do can men remain free agents, only by constant willingness to change their minds and start again can they be trusted with so great power as that to begin something new.

Thus, mutual release from past acts enables the possibility of *natality*, the birth of new ways of being and acting, new futures and ways upon which 'our hope relies' (Arendt 2006 [1961]: 189.

Arendt makes (1974 [1958]: 238) the observation that:

> The discoverer of the role of forgiveness in the realm of human affairs was Jesus of Nazareth. . . . The fact that he made this discovery in a religious context and articulated it in religious language is no reason to take it any less seriously in the strictly secular sense.

For Arendt, political perspectives on the significant discovery made by Jesus of Nazareth was first that:

> It is not true that only God has the power to forgive and second that this power does not derive from God – as though God, not men, would forgive through the medium of human beings – but on the contrary must be mobilised by men towards each other before they can hope to be forgiven by God also. Jesus' formulation is even more radical. Man in the gospel is not supposed to forgive because God forgives and he must do 'likewise', but 'if ye from you heart forgive', God shall do 'likewise'.
>
> (239)[8]

Forgiveness, then, arises between people. It is a *horizontal* relationship.

Arendt also explores the types of transgressions committed by mankind. She argues that mostly this is not extreme evil;[9] rather it is 'trespassing' and states that this 'needs forgiving, dismissing, in order to make it possible for life to go on by constantly releasing men from what they have done unknowingly' (240).

The power to forgive stands in contrast to vengeance, which is its exact opposite. For Arendt, it is vengeance which ensures that 'far from putting an end to the consequences of the first deed, every party remains bound to the process, permitting the chain reaction contained in every action to take its unhindered course' (240). As noted earlier, Arendt recognises that whilst vengeance is a predictable response to a wrong deed, forgiveness enables man to act in unexpected ways and thus express their freedom – a democratic move – since it:

> is the only reaction which does not merely re-act, it acts anew and unex-pectedly, unconditioned by the act that provoked it and therefore freeing from its consequences both the one who forgives and the one who is forgiven.
>
> (241)

A further key aspect of Arendt's thinking on forgiveness is the possibility of forgiveness for the sake of 'who' a person is rather than focusing on what they have done. Arendt notes that in religious thought, such as in Christianity, love is identified as the driving force behind the action of forgiveness. Arendt draws instead on a concept of respect built on Aristotelian *philia politikē*. She explains this as:

> A kind of 'friendship' without intimacy and without closeness; it is regard for the person from the distance which the space of the world puts between us, and this regard is independent of qualities which we may admire or of achievements which we may highly esteem.
>
> (243)

In this line of thinking, it is a mistake – a mistake which is often a feature of the modern world – to think we can only respect people whom we admire or like. Moreover, respect, understood in this sense of *philia politikē*, is important for forgiveness. Since such respect based on *philia politikē* 'concerns only the person, it is quite sufficient to prompt forgiving of what a person did for the sake of the person' (243).

Remembrance and forgiveness have become important themes in the political sphere since Arendt's death in 1975. I highlight two very different approaches: the first from Spain and the second from South Africa. These can be explored as 'case studies' in educational settings both to illuminate Arendt's theoretical ideas on forgiveness and to encourage students to explore how these could help them both in their own lives and in situations in the broader con-texts in which they live where there are different historical and current social issues and injustices to examine and address. In Appendix 1, I discuss ways that the case studies shared in this and subsequent chapters can be used as starting points.

Two case studies: differing approaches to the role of forgiveness in the public realm

Arising from his 'victory' in the Spanish Civil War, the dictator General Franco ruled Spain from 1938/9 to 1975. Mallinder (2008: 52) comments that:

> From the early 1970's when Franco's death seemed imminent, opposition parties formed a broad coalition to demand a clean break with the old system, political reform and a complete political amnesty.

This led, after Franco's death in 1975, to the law *Ley de amnistia* 46/1977, which became popularly known as the *Pact to Forget* (Tremlett 2006, Ash 1998, Biggar 2007). The justification was that an amnesty from 'everyone to everyone' was justified since both sides had committed 'bloody crimes' (Mallinder 2008: 67).

Two young Spaniards, Bea and Pablo describe how under the 'Pact to Forget':

> The post-war period and Franco's dictatorship became a taboo when the democracy arrived in 1975: Politicians thought that forgetting the past would help to avoid the ghosts of the dictatorship. And that interest of 'brooming' our past (for most of us our history) affected two very important areas: education and politics.
>
> (Chave 2012: 23)

Recognising that to move into the future with confidence depends on confronting and dealing with the past, a political movement began in 2000 to explore how to treat the past differently. In 2004, a commission was established, and in 2007, the *Law of Historical Memory* was passed. This enabled a process of investigation, rehabilitation and the healing of past wounds, enabling acts of forgiveness to take place in the public domain. This reached into educational settings through exploration of new understandings of the country's history; the importance of remembrance and the possibilities opened by a horizontal forgiveness between people. This included how such forgiveness could be understood as democratic – an expression of freedom to act in unexpected ways since the expected response is vengeance.

The approach to the past in South Africa in the 1990s had a strong focus on the role of forgiveness from the outset. The official end of Apartheid, and the subsequent elections in 1994, won by the ANC party with Nelson Mandela as the President, saw the creation of the Truth and Reconciliation Commission (TRC). Young-Bruehl (2006), drawing on Archbishop Desmond Tutu's account, *No Future without Forgiveness* (1999), explores both the centrality of forgiveness in the process of reconciliation and also who was able to forgive and receive forgiveness in this process. The TRC was not a court issuing sentences and punishments. Instead, the TRC was able to offer amnesties to perpetrators of violence who committed themselves to the following process. First, they had to fully disclose their actions, and the burden of demonstrating full disclosure in the public domain fell on them. This approach was adopted as it was felt that 'this brought out much more of the truth of what had happened than had the criminal trials conducted as apartheid ended (and while Afrikaners dominated the courts)' (Young-Bruehl 2006: 115). Then, once all the amnesty information had been assembled, the TRC organised meetings between 'perpetrators and their victims, people who had been tortured or lost family members and comrades either to the state policy of killing and torture or to the anti-apartheid resistance' (115). The meetings were organised so that 'the offenders spoke first; the victims could then tell their own stories and question the offenders' (115). It was after this that 'victims could decide for themselves whether to forgive or reconcile themselves with their offenders'. Forgiveness, it was understood, 'could not be requested of the victims by the Commission: it had to be freely chosen by the individuals who had been wronged' (115). Forgiveness was a public act by both parties. Moreover, it was understood that forgiveness could only be given by the victim of the act needing forgiving – both key features of Arendt's concept of forgiveness.

The process also highlighted the importance placed on forgiveness *to make a new way forward possible*. In this regard, Young-Bruehl comments that she was particularly struck by the story of two victims, widows of two black policemen assassinated by Eugene de Kock, former head of the South African security police's death squad. After listening to de Kock's remorseful account, Mrs Faku, one of the widows, said:

> I couldn't control my tears. I could hear him but I was overwhelmed by emotion, and I was just nodding as a way of saying, yes I forgive you. . . . I would like to hold him by the hand and say there is a future and that he can still change.
>
> (116)

Young-Bruehl comments:

> Mrs Faku spontaneously included in her act of forgiveness an invocation of the new beginning, the different future that a releasing act of forgiveness can make possible – and that repentance prepares the way for. She wanted his change of heart, his repentance, to be the beginning of his re-entry into the human fold he had denied and stepped out of.
>
> (116)

Exploring the aspect of forgiveness in education

Education can be a place to explore that the forgiveness proposed by Arendt is an *immanent* ethical response emerging through specific encounters with others. Discussions could include how Arendt's ideas differ from other ideas on forgiveness; the implications of this (which will vary in different educational contexts) and how these ideas respond to problems created by emergence through offering a release from past wrongdoings and opening possibilities for new futures.

Education can be a place for educators and students to examine how forgiveness, for Arendt, is one aspect of a whole. Receiving forgiveness is not a passive action on the part of the recipient. It requires *shuv* – a turn – away from one's past acts and a commitment, a mutual promise made publicly by both the one forgiving and the one asking for forgiveness, to redress wrongs and act in new ways. Mutual promising is explored later in this chapter.

The two cases from Spain and South Africa can be explored in educational settings as powerful examples of the way attitudes towards engaging with past wrongs have changed in recent history, highlighting the value of forgiveness in the process and its centrality for making the new a possibility. They draw attention to the importance in Arendt's thinking that the power to forgive arises between people speaking and acting with others. It is *a horizontal power* rather than an understanding of power as coming from 'above' – from some kind of divine or sovereign force. The two case studies present opportunities to discuss the importance of and relationship between both remembrance and forgiveness, including how attitudes have changed towards the need for remembrance – a remembrance without vengeance. Truth and Reconciliation Commissions, or approaches inspired by these, are now being instigated or considered in other countries. Students can also research and reflect on these contemporary examples, including how they relate to their own situation, as well as how and why such approaches can be problematic.

Before moving on to the role of mutual promising in Arendt's thinking, it is necessary to consider sensitively with students, in age-appropriate ways, the challenging issue of whether there are any situations where the wrongdoing committed is *beyond* forgiveness. An approach to ethics which proposes a crucial role for forgiveness needs to take seriously possible limits to such forgiveness, as well as examining the causes of extreme evil, if one wants to keep open the possibility that particular acts of extreme evil could be avoided in the future. It is therefore to this issue I now turn. The issue of extreme evil and the examples included are undoubtedly demanding, but it is important not to underestimate the capacity of students to examine and respond to them. Educators can explore ideas with students, through reflection, discussion and also aesthetic responses selected by the students themselves, for example via the media of prose, poetry, art, dance and music (see Appendix 1 for further guidance).

Limits to forgiveness – the issue of extreme evil

Young-Bruehl explains how Arendt's arguments regarding possible limits to forgiveness draw on an awareness of Kant's notion of radical evil, a type of evil 'which is rooted in (has its *radix* in) an evil motivation, an intention to do evil, a person's evil heart' (Young-Bruehl 2006: 2). Such radical evil is identified as being a rare occurrence by Kant and quite different from evil done out of ignorance or emerging through actions gone awry.

Arendt uses the term 'extreme evil' rather than 'radical evil'. She identifies the deliberate use of scientific methods to plan and carry out the extermination of the Jewish people as an instance of extreme evil. For Arendt, the Holocaust falls into a category *beyond the realm of human action* (a category where humans act in ways that go outside what it is to be and act as human) and therefore the acts committed there *cannot* fall into the realm of *human forgiveness*. Arendt was concerned to explore both the processes by which such a move *beyond the realm of the human* occurs and also the source of such extreme evil. Engaging with her ideas is important, including in education, both to understand better and respond to the Holocaust and other past acts of extreme evil and also to examine and respond to ways that extreme evils continue today.

In *Origins of Totalitarianism* (1973 [1951]), she identifies that in committing acts of extreme evil beyond the realm of the human, the perpetrators, through a perverted three-step process, remove the rights of a person or group of people and the extreme evil of totalitarianism emerges – a central concern in Arendt's writing. Through this perverted process a person or group appear to become 'superfluous' and are thus destroyable. First, judicial rights are removed. Second, the ability to make moral decisions is taken away, for example when, in Germany in the 1940s, a Jewish community leader could be forced to choose between 'betraying and murdering his friends or sending his wife and child for whom he is in every sense responsible, to their death; and even suicide would mean the immediate murder of his own family' (Arendt 1968 [1951]: 150). This leads to the third step in the creation of a person as superfluous: the removal of their ability to act spontaneously, the denial of any possibility of natality. There is an interesting link here to the experience of Frankl (2004 [1957]) discussed in Chapter 5. He argues, from his experience of totalitarian treatment in the Holocaust, that it was his and other prisoners' mental ability to resist this third step, the denial by their persecutors of their personhood and possibility for natality – the doing of a new, unexpected thing – that contributed to their survival. He gives as an example the choice made by some prisoners to share their very limited food with those weaker than themselves. It was these prisoners who survived better (even than some physically stronger prisoners) as the choice they were still able to make built up their mental resistance: resistance arising from the possibility they found, even in the most desperate situation, to claim their humanity, make moral choices and express their freedom to respond spontaneously to their extreme persecution in unexpected ways.

Recognising and understanding processes through which creating a person or group of people as 'superfluous' occurs – a form of dehumanisation – continues to be a very important area for study at a time when the world is grappling with how to move forward in new ways to be together. Arendt's thinking on this issue can support such exploration.

Another area examined by Arendt is the *source* of extreme evil. In her thinking, Arendt departs from Kant, who identified 'radical evil' as arising from an evil root within a person's heart, evil as the actions deliberately performed by an 'evil monster'. Arendt's controversial move was to argue for an understanding of the source of extreme evil, which is perhaps even more shocking – that it is possible to perform extreme evil *thoughtlessly*. In *The Human Condition* Arendt points out that:

> thoughtlessness: the heedless recklessness or hopeless confusion or complacent repetition of 'truths' which have become trivial and empty-seems to me among the outstanding characteristics of our time.
>
> (Arendt 1974 [1958]: Prologue p. 5)

Her thinking in *The Human Condition* and her experience in 1963 as the reporter for *The New Yorker* magazine at the trial of Eichmann, an SS Officer responsible, amongst other crimes, for the deportation of Jewish people to extermination camps, led to the development of her controversial idea of 'the *banality* of evil'. In a letter (cited in Bernstein 2002: 218) to her fierce critic, Gershom Scholem, she defends her position on the banality of evil, stating:

> It is indeed my opinion now that evil is never radical, that it is only extreme, and that it possesses neither depth nor any demonic dimension. It can overgrow and lay waste to the whole world because it spreads like a fungus on the surface. It is 'thought defying' as I said before because thought tries to reach some depth, to go to the roots, and the moment it concerns itself with evil, it is frustrated because there is nothing. That is its banality.

It is very important to note, however, that for Arendt, *banal evil is still extreme evil*, occurring in a realm which has created a person or group of people as superfluous. This point was misunderstood by many when Arendt first used the term 'the banality of evil' and it was popularised in the media. Arendt was criticised, even pilloried, in the ensuing controversy.[10] Arendt was *not* claiming that the evil committed was trivial. What *was banal* was the thoughtless way that the extreme evil was performed. When she looked at Eichmann in the court she had expected to come face to face with an evil monster: evil going to the very roots of his being, However, she found the reality very different and much more complicated.

For Arendt, actions of extreme evil place its perpetrators outside the realm of human forgiveness. Both Derrida and Young-Bruehl, however, explore whether there are other possible ways forward. Derrida (2001) proposes 'a "hyperbolic" ethical vision of forgiveness'. Janover (2005: 225) points out that 'hyperbolic' in the sense that Derrida is using it is not a rhetorical device meaning overblown but instead is 'hyperbolic' in its mathematical sense of eccentric or off-centred. Derrida's hyperbolic ethics 'depicts forgiveness as a moment of quandary and decision' (225) that 'should not be normal, normative, normalising. It should remain exceptional and extraordinary in the face of the impossible: as if it interrupted the ordinary course of historical temporality' (Derrida 2001: 45 cited in Janover 2005: 225). Janover proposes that Derrida's 'unconditional forgiveness' (which Janover calls 'radical forgiveness') 'severs or exceeds the loop that binds ordinary forgiveness to remorse, a change of heart, reconciliation, repatriation or restitution to the injured' (225). It is the interruption or severing of the ordinary loop of forgiveness which enables a new starting point, a new way to build the future even in the face of the most extreme evil.

Young-Bruehl (2006: 122), through her exploration of the Truth and Reconciliation Commission (TRC) in South Africa, argues that Arendt underestimates the human power to forgive. Young-Bruehl notes that Tutu and the TRC draw both on the Christian notion of forgiveness and the African notion of *Ubuntu*. Young-Bruehl sees parallels between Arendt's interpretation of Aristotle's *philia politikē* and *Ubuntu*. Tutu (1999: 34) explains *Ubuntu* in the following way:

> *Ubuntu* (in the Nguni group of languages) is very difficult to translate into a Western language. It speaks at the very essence of being human . . . you are generous, you are hospitable, you are friendly you are compassionate. You share what you have as if to say "my humanity is inextricably bound up in yours". We belong to a bundle of life. . . . To forgive is not just to be altruistic. It is the best form of self- interest. What dehumanises you inexorably dehumanises me. . . . Forgiveness gives people resilience, enabling them to survive and emerge still human despite all efforts to dehumanise them.

Young-Bruehl points out that the activities of the TRC were not part of a judicial process of punishment, and it is important not to make easy comparisons between the TRC and court processes. She concurs with Arendt that judicial processes are necessary to consider punishment, however inadequate they may seem, in the face of crimes against humanity. The processes are needed to establish the nature of such crimes and set precedents which can act as the grounds and guides for future legislation, for example the legislation enacted in the International Courts for the trial of crimes against humanity and genocide. However, what Young-Bruehl (2006: 121) does point out is that forgiveness, and the *Ubuntu* or *philia politikē* that makes forgiveness possible:

> needs to be cultivated by and in political processes where it is assumed (in Arendt's manner) that forgiveness, as a structural element of human affairs, is a necessity of political life. Adapted to local circumstances, forums like the TRC should be as much part of political life to deal with past conflicts as forums for treaty negotiation – promise making – are to secure against

future conflicts. Forgiveness is not just an action that can take place, it is an action which should be encouraged; it is not a process, not a replacement, for judicial process or punishment – it is not a way of encouraging criminals to act with impunity or a sense that all will be forgiven – but a potential means of preventing an endless imprisonment in the past.

For Young-Bruehl, the possibility of forgiveness is something which the TRC shows should not be underestimated: a *possibility* of forgiveness exists as long as humans exist in the world.

Exploring in education forgiveness and the issue of extreme evil

The concepts presented here on 'extreme evil' are undoubtedly theoretical but do have practical application in education, as the examples given demonstrate. Students of all ages are confronted with news of events many would deem extreme evil. This impacts on students whether education wants to engage with the issues or not. This is not to deny the importance of education as *skholé*, a place of protection. What teachers can do is encourage a safe space to explore the different ways one can respond to extreme evil in unexpected ways – an expression of freedom and thus a democratic move. It opens possible new futures closed down by the more expected response of vengeance which keeps us bound to the past in an endless cycle. Aesthetic activities, as suggested earlier in the chapter, can open ways to adopt the holistic learning approach to these challenging issues: an engagement which involves one's mind, physical senses and emotions (see Suzawa 2013 for further discussion). For example, teachers could draw on, as a *starting point*, the song "Don't Look Back In Anger" by the Manchester band Oasis. The song became an expression for many in Manchester and beyond of an alternative way to respond to the Manchester Arena bombing in 2017: an invitation to turn towards love and to the future rather than revenge.

Education can also be a place to explore the idea that extreme evil does not always arise from evil roots but can also arise from *banality*, from *thoughtlessness*, in relation to habits or 'repeated truths'. A deep questioning of such habits or 'truths' would reveal their 'wrongness', but often habits or ways of knowing and acting in the world remain unexamined, including by those in education. Such an argument can be expanded from Arendt's exploration of the 'banality of evil' demonstrated in the Holocaust to the contemporary issue of 'ecocide'.[11] As Orr (2004: 7) points out:

> The truth is that many things on which your future health and prosperity depend are in dire jeopardy: climate stability, the resilience and productivity of natural systems, the beauty of the natural world, and biological diversity.

It is worth noting that this is not the work of ignorant people. It is, rather, largely the result of work by people with BAs, BScs, LLBs, MBAs, and PhDs.

Orr (2004: 7) makes a connection between such thoughtlessness of contemporary ecocide and the comments of Elie Wiesel, writer and Holocaust survivor, at the Global Forum in Moscow in 1990, that despite extensive education received by the majority of Germans, heirs to the thinking of Kant and Goethe, this did not 'serve as an adequate barrier to barbarity' perpetrated during the Holocaust. For Wiesel, the problem with the education people had received was that

> It emphasized theories instead of values, concepts rather than human beings, abstraction rather than consciousness, answers instead of questions, ideology and efficiency rather than conscience.
>
> (Wiesel 1990 cited in Orr 2004 [1991]: 7–8)

Teachers and students can reflect on these ideas, both in relation to the Holocaust and genocides which continue to occur and also to extending Arendt's idea of committing extreme evil thoughtlessly to cover all participants in the wider natural world.

Examining the issue of extreme evil can never fully answer the question of where the limits lie beyond which humans cannot forgive, or indeed if there are such limits. This ambiguity is something to explore, reflect upon and discuss. Education and ambiguity are not always comfortable bedfellows, with education's tendency to 'sanitise knowledge' (Suzawa 2013:232) and to want to provide answers. Sitting with ambiguity, and the discomfort that lack of certainty can bring, is a further skill/attitude to be considered and developed in relation to forgiveness. When engaging with ambiguity, Suzawa (2013: 232) emphasises the possibilities opened by approaches which are not limited to analytical thinking and the reduction of thinking to 'very narrow or restricted modes of cognitive processes' where the mind 'operates along lines of logical reasoning or classification' and 'in terms of a reductivist strategy of knowledge building'. (This is an approach to the world which, as Little Bear [2000, 2011, 2016] highlights, dominates in Western thinking and to which Indigenous approaches offer other ways). For Suzawa, a cognitive revolution is needed, and is beginning to take place, to encourage the possibility of 'holistic modes of thought such as artistic creation'(232) and/or 'dialogical reasoning'. Ideas on engaging with these are discussed in Appendix 1.

Thus far in the chapter, I have focused on Arendt's thinking on forgiveness. However, it is important to remember that for Arendt forgiveness is not an isolated act. Central to her thinking on forgiveness is a turning away – a *shuv* – from past acts and a *commitment*, expressed through mutual promising, to new ways to be together. It is therefore to mutual promising that the chapter now turns.

The aspect of mutual promising in Arendt's thinking

Mutual promising

For Arendt, mutual promising is not just something that comes after forgiveness. Rather the act of promising as *shuv* – as a commitment to and action of turning from past wrongdoings – is integral to forgiveness. Forgiveness is not some kind of 'wishy-washy' thing; it involves change, which can include reparation for past wrongs and a commitment by all involved in the situation to new ways to be and be together in a shared world. It enables the possibility of a turn from, and a release from, the errors of the past arising from the irreversibility, unexpectedness and unboundedness of emergence towards the hope of new futures. Arendt highlights how the power to make and keep mutual promises is a horizontal power arising between people and that such power can be understood as dynamic, *as potentiality to act*. Whilst the future is unpredictable this unpredictability is a necessary part of political life. Unpredictability is the price of freedom to act and of not being ruled over by another. Arendt comments that this unpredictability is of a two-fold nature that:

> Arises simultaneously out of the darkness of the human heart, that is the basic unreliability of men who can never guarantee today what they will be tomorrow, and out of the impossibility of foretelling the consequences of an act in a community of equals[12] where everybody has the same capacity to act.
>
> (1974 [1958]: 244)

For Arendt, mutual promising has the potential to be a stabilising force which partially dispels the unpredictability that the future brings. Mutual promising can provide 'guideposts of reliability' and 'isolated islands of certainty . . . in a sea of uncertainty' (244). However, mutual promising loses 'its quality of the freedom to act' the moment that it is 'misused to cover the whole ground of the future and to map out the path secured in all directions' (244). If misused in this way, mutual promises 'lose their binding power and become self-defeating' (244), emphasising the importance of keeping open the possibility of the emergence of new and unexpected ways of knowing, being and acting in the world. As Arendt herself highlights, the idea of 'isolated islands of certainty' needs some careful consideration if emergence of the new is to be a possibility. In a radically unknown future, nothing can be 'certain'. However, one can make a promise to approach an issue with a *particular commitment* as circumstances emerge and change. For example, a promise to feed a child three times a day and provide the calories recommended for their age group would be a specific promise but too inflexible. It would close down the potentiality of making choices as different needs, hopes and desires emerge in the future: choices about the kinds of food the child wants to eat informed by the processes involved in food production, personal tastes and the possibility of moments of joyful shared indulgence. A different kind of promise is needed. Approaching 'the future as a site in which novelty is possible' (Facer 2016: 62), one can make a commitment to respond to emerging nutritional needs and choices of the child as they arise. The promise to do this can provide a child with a *secure* base, an 'isolated island of certainty', from which to explore the world and the issue of nutrition.

A significant feature of Arendt's thinking on mutual promising (1974 [1958]) is that she has *confidence* in the power people have to make mutual promises. For Arendt, the power to make *mutual* promises arises horizontally between people speaking and acting in 'spaces of appearance'. It provides an alternative to both domination of the self by another and also to isolation from the other. In *On Revolution* (1973 [1963]), Arendt draws attention to the importance of *mutuality* and the horizontal source of power in mutual promising. She cites the Mayflower Compact, declared by the colonists in 1620 as they approached America:

> Solemnly and mutually in the presence of God and one another we covenant and combine ourselves together in a civil Body Politic . . . and by virtue hereof enact, constitute and frame, such just and equal Laws, Ordinances, Acts, Constitutions, and offices, from time to time, as shall be thought most meet and convenient for the general Good of the Colony; unto which we promise all due Submission and Obedience.
>
> (Arendt 1973 [1963]: 173)

This is a commitment to respond to emerging situations 'as shall be thought most meet' (173) – and in such an emergent sense, the commitment to act in this way is an 'isolated island of certainty in a sea of uncertainty' (244).

Parekh (2007) identifies several key features which Arendt noted made this Compact distinctive. First, the Compact was a mutual promise, based on reciprocity and a presupposition of equality between people, to 'gather together the isolated power of individuals' into a body politic. This was *an alliance* rather than an agreement where individuals consented to give up their potential individual power in order to be ruled by another. Second, the Compact recognises that mutual promising necessarily takes place in the *presence of others* in plurality and emerges through an immanent process. It is *in and through the process of 'making public/publicity'* (Masschelein 2011, Masschelein and Simons 2013) that ideas and promises emerge through their declaration rather than preceding it. This focus on immanence resonates with the feminist ethics of care (explored in Chapter 3), which emphasises immanent ethics arising between people speaking and acting together rather than through reference to universal principles. As highlighted by Gilligan (see Ethics of Care 1984), such immanence is a strength, not a deficit position as portrayed in Kohlberg's model of stages of moral development. Third, in drawing up the Compact, the colonists showed remarkable *confidence in their capacity to make and keep promises to each other*. Although they made the promises in the presence of God, the promises were mutual. Parekh (2007: 75) notes that in Arendt's analysis the success of the Compact and subsequent compacts (in, for example, the American Revolution) was that the American political experience recognised that, whatever a person's past and motivation, 'People, in their singularity, could bind themselves into a community and thus human nature could be checked through common bonds and mutual promises'.

La Caze (2014) identifies two important differences between Arendt's concept of mutual promising, as seen in the Mayflower Compact, and social contract theory. First, the concept of promising that Arendt develops relates to **events** not theories. For Arendt, mutual promising is an *event* rather than a theory 'because it responds to specific difficulties in a particular context and is based on actual mutual confidence in the faithfulness and determination of others' (La Caze 2014: 217). Second, La Caze points out that Arendt's concept of promising differs from social contract theory in that Arendt's concept is based on a *genuine plurality*, a forming of *loose alliances* through which *the power to act* with unique others and keep promises is made possible. In contrast, social contract theory assumes a theoretical agreement between people whilst neglecting their differences. It assumes a willingness by individuals to sacrifice their power to act to an agreed state power. Another key feature of the *power enabling mutual promise-making* is that it is *a power potential*. Arendt (1974 [1958]: 200) points out that the word 'power' has its roots in the Latin *potentia*. Understood alongside *dunamis* – the Greek equivalent of the Latin word *potentia*, this foregrounds the meaning of power as a dynamic potential arising from free action: a potential for natality arising in and through 'spaces of appearance'. Hinchman and Hinchman (1994: 208) highlight how Habermas contrasts Arendt's conception of power as potentiality with that of Weber. Weber also conceives of power as potentiality, but it is the potentiality to force one's will upon another. Arendt, however, understands that the potential to exert one's power over another in the way Weber conceives is in fact violence and a desire for domination (see *On Violence (1970)* and *The Human Condition* 1974 [1958]).[13] She highlights the impact of non-violent protest or direct action, such as that proposed by Gandhi, as an illustration that public action has a power-potential even in the face of violence. Even death from violent acts demonstrates a power held by an individual. Death denies power to the person using such violence to dominate the person who has died (although this may be only a limited comfort to that person and those they are close to). As Frankl (2004 [1957]) identifies, the experience of enacting power by choosing how to respond to the violence exercised against him and other prisoners in the Holocaust illustrates the potential to have power even in the seemingly most powerless of situations. Indeed, it is this holding on to, and enacting of, this potentiality which Frankl believes maintained the humanity of the prisoners in the concentration camps and helped them to survive.

Habermas (1994: 244), however, argues that Arendt, in her focus on power emerging in communicative acts between people making mutual promises, ignores the structural rather than physical 'violence' which can block partici-pation in communicative acts in the public realm (i.e. not everyone has equal access to this realm). He argues that 'structural violence does not manifest itself as a force; rather it blocks those communications in which convictions effec-tive for legitimisation are formed and passed on (244)'. This is an important criticism, but engaging Andreotti's (2016) HEADS UP model to identify and challenge hidden barriers and assumptions, and drawing on Rancière's (2010a) concept of *dissensus*, alongside Arendt's natality, can help here (see Chapter 5 for an introduction to these models/concepts). Even those who, in the existing 'arrangement of the sensible', such as the young people involved in the Climate Strike Movement, have no right to speak can find ways to challenge existing structures and insert new ways of knowing, being and acting in the world. Schaap (2011) and Dikeç (2013) both explore the example of the *sans papiers* (immigrant workers in France who had no right to interrupt and rearrange the existing structures). These *sans papiers* claimed the right to appear and speak in the public realm, to have a 'voice' rather than merely produce 'noise' even though structural forces appeared to prevent this. These examples show there *is* the potential to challenge, to interrupt structural as well as physical violence and barriers to participation in speaking and acting in the public realm.

The role of mutual promising in keeping open the 'fleeting' power potential generated in 'spaces of appearance'

Arendt introduces a further important aspect of the role of promising and its relationship with power/potentiality. In the same way that 'spaces of appear-ance' are fleeting and only exist when people gather to speak and act together, so too is the power generated there. Power 'disappears the moment they depart' (Arendt 1974 [1958]: 244). Mutual promising, which the power generated in 'spaces of appearance' makes possible, has the important role of also then keep-ing that power open and alive. Mutual promising keeps open the possibility that a group of people can 'be bound together for an agreed purpose for which alone the promises are valid' (245). Such a binding together is superior both to approaches based on 'a will which somehow magically inspires them all' (245), or a world in which people are unbound by any promises or shared purpose. Keenan (2003: 86) endorses Arendt's approach, pointing out that:

> The force of promise lies in its ability to form a new political community or 'space' where none had been before, by deliberately leaving a trace or mark in the present that immediately becomes the past on whose basis the identity and the freedom of the community can be measured.

Keenan argues that promising can thus 'maintain or even enlarge' (88) the human power/potentiality for natality, for bringing in the new. For Arendt (1974 [1958]: 247), it is this potential for natality which brings hope. An example of such a 'trace' can be seen in the decision of the UK Court of Appeal to overrule permission granted by the UK Government for a third runway at Heathrow Airport, London. The Court argued that 'ministers did not adequately take into account the government's commitments to tackle the climate crisis' (Carrington 2020: online) made at the United Nations Paris climate talks in 2015 (talks which are explored in the case study in the next section). Whilst this decision may still be overturned (and perhaps also responds to other political interests alongside environmental concerns), it nevertheless does open potential for optimism.

Thus far in this section I have discussed Arendt's ideas on mutual promising and the hope it has potential to bring into the world. Education in many different contexts can be a place to consider these ideas, as I now go on to explore.

Exploring mutual promising in education

The aspect of mutual promising proposed by Arendt presents opportunities for exploration across a variety of educational settings. Discussions and responses will vary in different settings, based on cultural and socio-economic differences and past histories, for example how trustworthy other countries have been in mutual promises made.

A place to explore the relationship between forgiveness and mutual promising

Education can be a place to explore *the relationship* between forgiveness and mutual promising and how this can contribute to new ways to be together in the world. The case studies on Spain and South Africa introduced earlier in this chapter can provide starting points for this exploration. Education can also be a place to examine the challenging issue of whether and how reparations for past and current injustices form part of *shuv* – the commitment to turning from wrongdoings towards new ways of acting central to Arendt's two-fold concept.

A place to explore how mutual promising can keep open the power/potentiality which speaking and acting can create – case study: the United Nations and climate change

Education can be a place to explore how mutual promising can keep open the power/potentiality which speaking and acting can create. The following case study on a United Nations climate initiative can be discussed in education, or students can be encouraged to research the area for themselves. Ideas for reflection can include the role and influence of mutual promising in the process of treaty-making; the 'fleeting' power potential (or lack of it) that such treaty-making can open up and ways that such power potential could be enhanced.

Different contexts will create very different perspectives and reflections. For example, students in countries in the Global South, which have contributed far less to global warming but stand to be much more adversely affected by it, will bring different perspectives from students situated in the Global North. They will also bring different experiences of confidence in promise making between their own countries and those of Global North. These different viewpoints can also be introduced by teachers in the Global North to their students. The film *Thank you for the rain* (2017), produced as part of a collaboration between the Kenyan farmer and climate activist Kisilu Musya and the Norwegian film-maker Julia Dahr, could be a helpful resource here. Challenging hegemonic processes, Musya insisted on a fully collaborative project, including doing his own filming, rather than be the 'topic' of research as was first mooted. The film brings very different perceptions on the impacts of climate change, what being a climate activist means in different contexts and very different perspectives on the negotiations in the COPD 21 in Paris in 2015, in which Musya participated.

Case Study: In 1992, at the Earth Summit in Rio de Janeiro, over 190 participating governments signed The United Nations Framework Convention on Climate Change (UNFCCC). The long term aims of the treaty are to avoid dangerous human interference with the climate system and guide the global commitment by all countries to 'mitigate' climate change through reducing greenhouse gas emissions. An annual conference (known as the Conference of the Parties or COP) of all governments who ratified the treaty is held to set targets and review progress towards aims. Each country has a vote, and whilst smaller countries do not always have the financial capacity to bring a large number of negotiators to influence what is voted upon, they can and do group together to pool resources and influence. They also receive support and training in international negotiating from the United Nations. In some years, the conferences take the form of higher profile 'summits' which become known by the cities where they were held. For example, the Kyoto protocol was established in 1997 at the Kyoto Summit in Japan.

Since the first conference there has been growing scientific evidence of the negative effects of global warming, and humans' role in it (for example see IPCC 2007, 2014).[14] At COP17, held in Durban in 2011, nations agreed to establish by 2015, new, legally binding agreements to limit global warming. Subsequently, at the 2015 summit in Paris, commitments were made to form an *alliance* to reduce carbon emissions to 1.5C. This took many by surprise, after so many failed negotiations in the past which were very general and unsuccessful. The emphasis on *alliances* to work towards a target, rather than a handing over of power to another body, echoes the importance placed by Arendt on the potentiality of mutual promising, as in the *Mayflower Compact*, (see discussion on page xx). The case study, however, does highlight that promise making requires optimism: 'actual mutual confidence in the faithfulness and determination of others' to keep promises made (La Caze 2014: 217).

It was initially heartening that China and the USA ratified the agreement in September 2015, citing the importance of setting an example to encourage other nations and thus providing some cause for optimism. However, the USA then withdrew its support for the agreement in 2017, following the election of Donald Trump. The 24th meeting of COP took place in Katowice, Poland, in December 2018 to operationalise the Paris agreement, and it was successful, albeit at the last minute, in achieving this aim. The agreement sets out how different countries will:

> provide information about their Nationally Determined Contributions, the plans developed by each country that describe their domestic climate actions and . . . mitigation and adaptation measures as well as details of financial support for climate action in developing countries,
>
> (UN, online)

A global stock-take of the effectiveness of climate action will take place in 2023 to assess progress made in the development and transfer of technology to assist carbon reduction.

As of January 2019, 184 parties had ratified the Paris Agreement (Centre for Climate and Energy Solutions, 2019). Such mutual promising can attempt to forge the possibilities of new communities of action. Such communities are vulnerable to political change and lack of adherence to the mutual promises made. Meetings of COP will continue with mixed success. However, as Arendt reminds us, it is the human potential for natality 'which ultimately can be "the miracle" which saves the world' (Arendt 1974 [1958]: 247).

Education as a place to encourage the role of shuv – case study: Red*thread*

As well as thinking about ideas on mutual promising, education can also be a place to *enact* the ethical responsibility to encourage the *shuv* central to Arendt's concept of forgiveness and mutual promising. Such an understanding of *shuv* does not focus on what has been done (although this informs the promises made). Instead *shuv* focuses on what needs to be done now as well as commitment to adapt in the future to keep alive the commitment which *shuv* makes possible. This is a challenging demand – recognising that vengeance, with its focus on past injury – is a more common human response. However, it is a 'turning away/turning towards' which *teshuvah* ethically demands of us since it is vengeance which ensures that:

> Far from putting an end to the consequences of the first deed, every party remains bound to the process, permitting the chain reaction contained in every action its unhindered course.
>
> (Arendt 1974 [1958]: 240)

Turning to different ways is hard and requires courage. Education can be a place both to discuss the challenges of turning to new ways of being and acting as well as a place to provide support to students to meet these challenges. The *Red*thread case study contains examples of support offered to students and these can be used as a *starting point* for discussion and activities with, and support to, students of different ages in their own particular educational setting.

*Red*thread, a charity established in London UK in 1995, supports young people in South London, UK through 'the vulnerable transition of adolescence' by providing a range of services including a youth club, young people's clinic and wellness centre and a youth violence intervention scheme based in two London NHS hospitals. The name *Red*thread comes from Greek mythology,[15] which tells of a labyrinth (maze) guarded by a minotaur and famous for being unsolvable. Every year seven young women and seven young men were sacrificed to the maze. However, when it was Theseus' turn to enter and be sacrificed to the minotaur, the Cretan princess Ariadne helped him by giving him a red thread which he trailed behind him as he went through the maze. He defeated the minotaur and then used the thread to find his way out. *Red*thread comment, 'We love the essence of this myth: of someone helping someone else to face and solve their challenges' (*Red*thread.org: online). On their website, *Red*thread provides case study examples of its work which ranges from confidence-building and tackling bullying to finding practical and emotional ways to move on from gang participation and the cycle of vengeance which contributes to knife crime, a significant and worsening issue in London, UK. The breadth of support offered by *Red*thread provides a variety of 'entry-points' for discussion with different age groups and in different settings on how the act of *shuv* explored in this chapter is not always easy and how support for such actions of turning to new ways is needed and can be accessed in their own particular context. Students can explore these examples from *Redthread* and, if circumstances are appropriate, also reflect on experiences of their own. They can consider ways that forgiveness and mutual promising can be enacted in their own lives in meaningful ways, thereby opening new futures both for themselves as individuals and for the wider context in which they live.

Ethics of care, ethical relationality and Arendt's thinking on forgiveness and mutual promising: drawing some threads together

Throughout this chapter I have made reference to possible connections between Arendt's *immanent approach to ethics* emerging horizontally in events and feminist ethics of care as well as Indigenous thinking foregrounding ethical relationality. This is not to present Arendt as a feminist care ethicist, or a 'precursor' to such ideas. Rather, the aim is to explore what bringing the ideas of feminist thinkers, and also those of indigenous thinkers into conversation with Arendt's ideas, opens up. For example, Arendt argues that forgiveness and mutual promising opens up possibilities for inserting ourselves in new ways into the 'web of relations in the world' (Arendt 1974 [1958]:

184). Developing an 'enlarged mentality' through the immanent Arendtian processes of visiting, imagination and understanding discussed in Chapter 5 opens opportunities to consider how that 'web of relations' could include both the human and other parts/participants in the wider natural world, as in the thinking of Braidotti (2006, 2013), Haraway (2008, 2015, 2016) and Tronto and Fisher (1991). This is a move beyond Arendt, but one which bringing ideas from different thinkers together can encourage. Exploring 'the web of relations in the world' further invites exploration of 'the inseparable relationship between humans and earth, inherent to Indigenous Peoples' (KARI-OCA 2 Declaration 2012: online), the ethical responsibilities this generates and the possibilities of enacting Arendt's conceptions of both *shuv* and *natality* in such an expanded 'web of relations'. These ideas are taken forward and explored further in Chapter 8.

Education, *potentia* and response-ability

Sterling (2010b: 217) points out that an education which highlights environmental and social issues is asking students to develop an 'expanded and ethical sense of concern/engagement' with these issues. However, if this is asked of students they also need to be encouraged to explore and develop ways that they can *actively* respond to these demands. This active response is what Sterling terms 'response-ability' (217). As discussed in this and earlier chapters this is challenging since in Western society, dominated by a sense of *potentas*, power is often reified as something outside of a person and [mainly] held by others.[16] This can lead to a lack of action, a lack of 'response-ability', since one feels one does not have the power to act. Engaging with Arendt's challenging notion of power as potentiality, and the role of forgiveness and mutual promising in opening this potentiality, can inform and promote 'response-ability'. In this framing, individuals, including students of all ages, reposition themselves as actors who can generate such *power/potential*.

Education can be a place to reflect on action and also to act: a place for the *dissensus* or interruption that forgiveness and mutual promising can open up, a place for the ethical *shuv* or turn that lies at the heart of the concept. Arendt's approach is built on working through and acting on specific ethical issues with others unlike ourselves, in specific contexts and events: immanence rather than universalism as a source of ethics. *Shuv* – turning – also emphasises the dynamic nature of ethics. *Shuv* can be seen as a moment of ethical choice, a moment of undecidability placed on every actor in the world, and one which opens the possibility of new ways to be together. As Smedes (1996: 171) reminds us:

> Forgiving does not erase the bitter past. A forgiven memory is not a deleted memory. Forgiving what we cannot forget creates a new way to remember. We change the memory of our past into a hope for our future.

I first saw this challenging and brave quotation from Smedes on a 'Thought for the Day' board at the Oval Underground station, London, UK, in the days after a terrorist attack. It stopped me in my tracks, and I exchanged words of both sorrow and hope with a fellow traveller who had also paused. The 'thought', the event of encountering, the greetings exchanged: all had a profound effect on me – and still do.

Events and encounters with others unlike ourselves are critical for encouraging the possibility of opening 'spaces of appearance': spaces which include the possibility of responding ethically to challenges which the irreversibility, unpredictability and unboundedness of new ways of being, knowing and acting bring into the world. The possibility of such encounters is therefore the focus of the final chapters of this book.

Notes

1 Plurality is understood as where there is opportunity to speak and act with others and where each is open to the stance the other expresses (see Chapter 4 for further discussion).
2 Toposki (2011) includes analysis of *teshuvah* in Jewish religious traditions but also explores how whilst both Levinas and Arendt approach it from within these religious groundings their translation and understanding of it differs from such understandings.
3 A term I introduced in Chapter 2 – see discussions in Anzaldúa (1999) and Mignolo and Tlostanova (2006).
4 Normative – conforming to or based on preconceived often universalised norms or ways of being and thinking.
5 This is a collection of Arendt's writings from the 1930s through to the 1960s.
6 The process in which one acts but 'one did not, and could not, have known what one has done' is what Arendt (1974 [1958]: 237) calls 'unboundedness'. Unboundedness arises because others take up our beginnings in unexpected ways we did not and could not have anticipated.
7 For further discussion of this notion of 'releasing' see Young-Bruehl 2006: 100.
8 Arendt cites Matthew 18:35 and Mark 11:25 in her footnote in *The Human Condition* (1974 [1958]: 239) to support this argument.
9 The issue of 'extreme evil' is explored later in this section.
10 Arendt's resistance to taking a simplistic approach to the issue, raised at the Eichmann trial, of the role of Jewish councils in Germany and German occupied countries (see Benhabib 2000) further amplified the controversy.
11 Ecocide – such extensive damage to, destruction of or loss of ecosystem(s) that the very survival of that ecosystem and all who form part of it is threatened. Whilst not yet recognised by the United Nations as an international crime, organisations such as STOPECOCIDE are 'campaigning for it to be recognised as an atrocity crime at the International Criminal Court – alongside Genocide, War Crimes and Crimes Against Humanity' (STOPECOCIDE: online).
12 Habermas (1994) questions whether there is an equality to act. I explore this issue later in the chapter in discussions around structural violence.
13 For example, see Arendt's discussions in *The Human Condition* (1974 [1958]: 244).
14 According to the Intergovernmental Panel on Climate Change (IPCC) (2007), there is a dual relationship between sustainable development and climate change. On the one hand, climate change influences key natural and human living conditions and thereby also the basis for social and economic development, whilst on the other hand, society's

priorities on sustainable development influence both the GHG emissions that are causing climate change and the vulnerability thus caused. Climate policies can be more effective when embedded consistently within broader strategies designed to make national and regional development paths more sustainable. This occurs because the impact of climate variability and change, climate policy responses, and associated socio-economic development will affect the ability of countries to achieve sustainable development goals. Conversely, the pursuit of those goals will in turn affect the opportunities for, and success of, climate policies. Research on climate change was published by the IPPC in its *5th Assessment Report (2014)*.

15 *Red*thread has significance in other cultures too. For example, in South East Asian cultural traditions an invisible red thread joins together those destined to meet: a symbol of connection and relationality.

16 See also discussions in Graeber 2001.

References

Andreotti, V. (2016) The educational challenge of imagining the world differently. *Canadian Journal of Development Studies Revue Canadienne d'études du développement*, 37(1): 101–112.

Anzaldúa, G.E. (1999) *Borderlands/La frontera: The new Mestiza*. San Francisco: Aunt Lute Books.

Arendt, H. (1968 [1951]) *Totalitarianism: The origins of totalitarianism: Part three*. San Diego, CA: Harcourt, Brace, Jovanovich.

Arendt, H. (1970) *On violence*. San Diego, CA: Harcourt, Brace, Jovanovich.

Arendt, H. (1973 [1951]) *The origins of totalitarianism*. Orlando: Harcourt.

Arendt, H. (1973 [1963]) *On revolution*. Harmondsworth: Penguin.

Arendt, H. (1974 [1958]) *The human condition*. Chicago, IL: University of Chicago Press.

Arendt, H. (1994 [1954]) Understanding and politics. In *Essays in understanding*. New York: Schocken Books.

Arendt, H. (2006 [1961]) Truth and politics. In *Between past and future: Eight exercises in political thought*. London: Penguin.

Arendt, H. (2007) *The Jewish writings*. New York: Schocken Books.

Ash, T.G. (1998) The truth about dictatorship. *The New York Review of Books*, 45(3).

Benhabib, S. (2000) Arendt's Eichmann in Jerusalem. In D. Villa (ed.) *The Cambridge companion to Hannah Arendt*. Cambridge: Cambridge University Press, pp. 65–85.

Bernstein, R.J. (2002) *Radical evil: A philosophical investigation*. New York: Wiley.

Biggar, N. (ed.) (2007) *Burying the past: Making peace and doing justice after civil conflict*. Washington, DC: Georgetown University Press.

Birmingham, P. (2006) *Hannah Arendt and human rights: The predicament of common responsibility* (Series: Studies in Continental thought). Bloomington: Indiana University Press.

Botbol-Baum, M. et Roviello, A. (co-ords.) (2013) *Arrachement et évasion: Levinas et Arendt face à l'histoire (Partie 1)*. Paris: Vrin.

Braidotti, R. (2006) The ethics of becoming-imperceptible. In C. Boundas (ed.) *Deleuze and philosophy*. Edinburgh: Edinburgh Press, pp. 133–159.

Braidotti, R. (2013) *The posthuman*. Boston, MA and Cambridge: Polity Press.

Carrington, D. (2020) Heathrow third runway ruled illegal over climate change. *The Guardian*, 27th February. Available at: www.theguardian.com/environment/2020/feb/27/heathrow-third-runway-ruled-illegal-over-climate-change [Accessed 8.4.2020].

Centre for Climate and Energy Solutions (C2SE) 2019 (*Paris Climate Agreements, Q&A*. Available from: https://www.c2es.org/content/paris-climate-agreement-qa [Accessed 27.3.2019].

Chave, J. (2012) *The story of the Basque children of 37: In the past, the present and into the future* (Unpublished dissertation submitted for BA Honours degree in Social Anthropology). Goldsmiths College, University of London, London.

Derrida, J. (2001) *On cosmopolitanism and forgiveness*. Dooley, M. (tr.). London: Routledge.

Dikeç, M. (2013) Beginners and equals: Political subjectivity in Arendt and Rancière. *Transactions of the Institute of British Geographers*. Royal Geographical Society (with the Institute of British Geographers), 38(1): 78–90.

Donald, D., Glanfield, F. and Sterenberg G. (2012) Living ethically within conflicts of colonial authority and relationality. *Journal of the Canadian Association for Curriculum Studies*, 10(1).

Ethics of Care (2011) *Gilligan*. Available at: https://ethicsofcare.org/carol-gilligan/ [Accessed 17.3.2020].

Facer, K. (2016) Using the future in education: Creating space for openness, hope and novelty. In H.E. Lees and N. Noddings (eds.) *The Palgrave international handbook of alternative education*. London: Palgrave Macmillan, pp. 53–78.

Frankl, V. (2004 [1957]) *Man's search for meaning: The classic tribute to hope from the Holocaust*. London: Rider.

Graeber, D. (2001) *Toward an anthropological theory of value: The false coin of our own dream*. New York: Palgrave Macmillan.

Habermas, J. (1994) Hannah Arendt's communication concept of power. In L.P. Hinchman and S. Hinchman (eds.) *Hannah Arendt: Critical essays*. Albany, NY: SUNY Press, pp. 221–230.

Haraway, D. (2008) *When species meet*. Minneapolis and London: University of Minnesota Press.

Haraway, D. (2015) Anthropocene, Capitalocene, Plantationocene, Chthulucene: Making kin. *Environmental Humanities*, 6: 159–165.

Haraway, D. (2016) *Staying with the trouble: Making kin in the Chthulucene*. Durham, NC: Duke University Press.

Hinchman, L.P. and Hinchman, S. (1994) Introduction to part four: Power and action. In L.P. Hinchman and S. Hinchman (eds.) *Hannah Arendt: Critical essays*. Albany, NY: SUNY Press, pp. 209–210.

Intergovernmental Panel on Climate Change (IPPC) (2007) *Climate change 2007: Working group III: The dual relationship between climate change and sustainable development*. Available at: www.ipcc.ch/publications_and_data/ar4/wg3/en/ch2s2-1-3.html [Accessed 12.9.2016].

Intergovernmental Panel on Climate Change (IPPC) (2014) *Fifth assessment report*. Available at: www.ipcc.ch/ [Accessed 16.6.2014].

Jakub, F. (2005) Political conditions of philosophy according to Arendt. In D. Gard, I. Main, M. Oliver and J. Wood (eds.) *Inquiries into past and present*, Vol. 17. Vienna: IWM Junior Visiting Fellows' Conference.

Janover, M. (2005) The limits of forgiveness and the ends of politics. *Journal of Intercultural Studies*, 226(3): 221–235.

Jonas, H. (1984) *The imperative of responsibility: In search of an ethic for the technological age*. Chicago, IL: The University of Chicago Press.

KARI-OCA 2 Declaration (2012) *KARI OCA 2 declaration: Indigenous peoples global conference on RIO + 20 and mother earth*. Accepted by Acclamation, Kari-Oka Village, at Sacred Kari-Oka Púku, Rio de Janeiro, Brazil, 17th June. Available at: https://wrm.org.uy/other-relevant-information/kari-oca-2-declaration-indigenous-peoples-global-conference-on-rio-20-and-mother-earth/ [Accessed 17.12.2019].

Keenan, A. (2003) *Democracy in question: Democratic openness in a time of political closure*. Stanford: Stanford University Press.

La Caze, M. (2014) Promising and forgiveness. In P. Hayden (ed.) *Hannah Arendt: Key concepts*. Durham, NC: Acumen, pp. 209–221.

Little Bear, L. (2000) Jagged world views colliding. In M. Batisse (ed.) *Reclaiming indigenous voice and vision*. Vancouver: University of British Columbia Press. Available at: http://www.learnalberta.ca/content/aswt/worldviews/documents/jagged_worldviews_colliding.pdf [Accessed 14.10.2019].

Little Bear, L. (2011) *Native science and western science: Possibilities for collaboration*. Lecture on 4th of March at Arizona State University. Available at: https://www.youtube.com/watch?v=ycQtQZ9y3lc [Accessed 15.11.2018].

Little Bear, L. (2016) *Blackfoot metaphysics 'waiting in the wings'*. Congress of the Humanities and Social Sciences Big Thinking Lecture, 1st June. Available at: https://www.youtube.com/watch?v=o_txPA8CiA4 [Accessed 18.11.2018].

Masschelein, J. (2011) *Experimentum scholae*: The world once more . . . but not (yet) finished. *Studies in the Philosophy of Education*, 30(5): 529–535. doi: 10.1007/s11217-011-9257-4

Masschelein, J. and Simons, M. (2013) *In defence of schools: A public issue*. Leuven: E-ducation, Culture and Society Publishers. Available at: http://ppw.kuleuven.be/ecs/les/in-defence-of-the-school/masschelein-maarten-simons-in-defence-of-the.html [Accessed 3.2.2016].

Mallinder, L. (2008) *Amnesty, human rights and political transitions: Bridging the peace and justice divide*. Oxford and Portland: Hart Publishing.

Mignolo, W.D. and Tlostanova, M.V. (2006) Theorizing from the borders: Shifting to geo- and body-politics of knowledge. *European Journal of Social Theory*, 9(2): 205–221.

Orr, D. (2004) *Earth in mind: On education, environment, and the human prospect*, 2nd Edition. Washington, DC: Island Press.

Parekh, S. (2007) *Hannah Arendt and the challenge of modernity*. London: Routledge.

Rancière, J. (2010a) Who is the subject of the rights of man? In J. Rancière (ed.) *Dissensus: On politics and aesthetics*. London: Continuum, pp. 70–83.

*Red*thread.org (online) *About us: Our name*. Available at: www.redthread.org.uk/about-us/ [Accessed 17.3.2020].

Schaap, A. (2011) Enacting the right to have rights: Jacques Rancière's critique of Hannah Arendt. *European Journal of Political Theory*, 10(1): 22–45. doi: 10.1177/1474885110386004

Smedes, L.B. (1996) *The art of forgiving*. New York: Random House.

Sterling, S. (2010b) Living in the earth: Towards an education for our time. *Journal of Education for Sustainable Development*, 4(2): 213–218. doi: 10.1177/097340821000400208

STOPECOCIDE (online) *Making ecocide a crime*. Available at: www.stopecocide.earth/making-ecocide-a-crime [Accessed 30.1.2020].

Suzawa, G. (2013) The learning teacher: Role of ambiguity in education. *Journal of Pedagogy*, 4(2): 220–236.

Thank you for the Rain (2017) *IMDbPro*. A film by Julia Dahr and Kisili Musya. Official website available at: https://thankyoufortherain.com/

Topolski, A. (2011) The ethics and politics of *Teshuvah*: Lessons from Emmanuel Levinas and Hannah Arendt. *The University of Toronto Journal of Jewish Thought*, 2. Available at: http://tjjt.cjs.utoronto.ca/wp-content/uploads/2013/11/Anya-Topolski-The-Ethics-and-Politics-of-Teshuvah-Lessons-from-Emmanuel-Levinas-and-Hannah-Arendt-JJT-Vol.-2.pdf [Accessed 4.11.2015].

Topolski, A. (2015) *Arendt, Levinas and a politics of relationality* (Reframing the boundaries: Thinking the political). London: Rowman & Littlefield International.

Tremlett, G. (2006) *Ghosts of Spain: Travels through a country's hidden past*. London: Faber and Faber Limited.

Tronto, J.C. and Fisher, B. (1991) *Toward a feminist theory of caring*. In E. Abel and M. Nelson (eds.) *Circles of care*. Albany, NY: SUNY Press, pp. 36–54.

Tutu, D. (1999) *No future without forgiveness*. New York: Doubleday.

Wiesel, E. (1990) *Remarks before the Global Forum*. Moscow: The Global Forum of Spiritual and Parliamentary Leaders on Human Survival.

Young-Bruehl, E. (2006) *Why Arendt matters*. New Haven: Yale University Press.

7 Intersubjective first-person encounters and encouraging the possibility of emergence of new subjectivities

Introduction

In this chapter I bring together ideas I have explored thus far to argue that intersubjective first-person encounters with other humans under conditions of plurality[1] open the possibility of a shift from a static or substantial concept of the human subject to the possibility of dynamic emergent subjectivity. In this line of thinking, the self emerges *in* and *through* radically open processes in 'the space between I and we' (Topolski 2015: 176) 'never outgrowing its open malleable ego boundaries' (Keller 1986 p. 134). Subjectivity arises *in* and *through* rather than preceding encounters between the self and the other. Intersubjective first-person encounters also have *potential* to be sources of immanent ethical responses between unique individuals. Through the process of opening spaces which 'respond to specific difficulties in a particular context' (La Caze 2014: 217), ethical responses can emerge in encounters 'based on actual mutual confidence in the faithfulness and determination of others' (217). Arendt's process of forgiveness and mutual promising explored in the previous chapter can help to support emergence of such ethical responses.

I include practical examples of how intersubjective first-person encounters can be encouraged in educational contexts. The case studies here and in Chapter 8 are deliberately drawn from *my own first-person/first-being encounters* in situations operating at the 'margins' of mainstream UK education. My own teaching life has been in these 'marginal places'. The practices developed in these 'marginal places' are designed to support students who are 'not doing well' academically and or emotionally in mainstream UK settings still strongly influenced by dominant Western (Eurocentric) framings which understand education as places to develop autonomous rational subjects. They provide insights beneficial to these students but equally to those in 'mainstream' settings. The examples I provide can also encourage recollection and reflection of educators' and students' own first-person encounters in the educational settings and contexts where they are located. In discussing these ideas during the research for this book, it was noticeable how everyone had their own first-person encounters and examples to share and reflect upon.

I conclude the chapter by raising the possibility of extending arguments made in relation to first-*person* intersubjective encounters to first-*being* encounters with the wider natural world and indicate that such first-being encounters are the focus of the next chapter.

Sterling (2010b: 214) recalls reading a curious notice on the bus he uses. It says, 'No eating or drinking on the bus'. He points out, however, that we travel 'in the bus' commenting:

> The difference in perspective that such experiences invoke is this: we are not on the Earth, we are in the Earth, we are inextricably actors in Earth's systems and flows, constantly affecting and being affected by everything natural and human, in dynamic relation (Metzner 1995). We are unavoidably participative beings. And yet, deeply embedded in the Western psyche, although we know participative reality to be true, there is a powerful operative myth of separateness.

Through exploring intersubjective encounters, I challenge this 'myth of separateness' – a (Western) Eurocentric worldview introduced and questioned in Chapters 2 and 3. Such challenging opens the possibility of emergence of new ways of being a human subject not restricted by notions of rationality, autonomy and separation between the self and the other.

Exploring ideas on *intersubjectivity* is a necessary starting point in order to think, at least briefly, about what is being referred to and understood in the term *intersubjective* in the phrase 'intersubjective encounters'. It is therefore to these ideas I now turn.

The development of ideas of intersubjectivity

Parekh (2007: 70) suggests Husserl as an interesting starting point for theorising the notion of intersubjectivity (i.e. ideas of subjectivity arising between/ inter subjects) and how this developed in different ways in nineteenth- and twentieth-century thinking. She comments that:

> For Husserl, since the basis of phenomenology was the self-constitution of the ego, the experience of other egos became important in order to avoid solipsism.[2] His concern was with what it meant to live in a world that was shared with others and thus had shared objects, language, meaning etc.

However, for Husserl, although the world has an objective existence outside of 'me'- a world in common, which is shared intersubjectively (between subjects) – it does not constitute 'who I am'. The self or ego is still constituted of an essence which one can reveal to the world.

In the twentieth century, some theorists began to conceptualise subjectivity in a different way. Rather than the subject preceding the intersubjective (what emerges in the encounter between two subjects), various understandings of intersubjectivity were developed in which the intersubjective precedes and informs the subject. Biesta (1994, 2006) draws attention to three thinkers who explore this possibility: Mead, Dewey and Habermas. Biesta comments that for Mead the whole of society is prior to the part an individual plays in it. Dewey and Habermas, however, emphasise the role that communication plays in building a common world. For Dewey, responding actively to the signals from another builds a world in common, and this informs *who* a person is. This is not a mechanical process, but rather one that emerges through co-operation.

In his thinking on intersubjectivity, Habermas (1984: 285–286) emphasises what he calls 'communicative action', contrasting it with 'strategic action'. In strategic action the individual is aiming to get agreement for a predetermined outcome, with the emphasis on pursuing one's own goals. In communicative action, however, agreement occurs whenever 'the actions of the agents involved are coordinated, not through egocentric calculations of success, but through the acts of reaching understanding'. Biesta (1994) points out that at first sight communicative action can appear to be merely a linguistic exchange, but that further study of Habermas indicates how he considers communicative action to be the birthplace of individual identity. Habermas states, for example, that 'identity formation takes place through the medium of linguistic action' (Habermas 1984: 58 cited in Biesta 1994: 310).

Ideas proposed by Mead, Dewey and Habermas can contribute to the possibility of a form of education which is not about bringing the pre-rational child to a state of rational autonomy or a pedagogical action considered to be a one-way process. Instead, pedagogical action can be conceived of as a co-constructed, intersubjective process through which meaning and *who* one is as a unique subject emerge. Such approaches acknowledge how both meaning and subjectivity cannot be produced in a predetermined way since individuals will respond differently to shared ideas and communication.

What is still problematic, however, in Dewey's and Habermas' thinking, is the idea of a common or shared understanding that can be reached through communication and negotiation. Mouffe (2007) challenges this idea with her thinking on agonistic pluralism (ideas explored in Chapter 5). Agonistic pluralism allows one to approach the other not as an enemy but instead as an adversary with whom one can argue without trying to erase them and their different points of view. In such an approach, rather than an emphasis on building a common world that everyone can agree on, the emphasis is on how to live in a world of difference and *uniqueness*: a world in which, to paraphrase Arendt (1994 [1954]: 322), 'bearing with' rather than erasing 'strangeness' is a possibility.

In Chapters 4 and 5 I argued for the importance of events or encounters in plurality – that is to say encounters with others unlike oneself where each is receptive to the expression of the other – to open spaces in and through which such *unique subjects* can appear and bring something new into the world. To expand my earlier argument I now draw on ideas of Arendt, Loidolt (2016), Keller (1986) and Braidotti (2011a, 2011b) to argue that such encounters under conditions of plurality can be understood as processes of *intersubjective encountering* and the contribution such processes can make to sustainable and democratic ways to be in the world together, including in education.

Intersubjective first-person encounters

Loidolt emphasises the importance of Arendt's phenomenological roots as the starting point for understanding how Arendt's notion of the emergence of *who* a person is is actualised in an event in plurality. Phenomenology, broadly understood, emphasises the study of the appearances of things (phenomena), or the ways we experience phenomena, as they appear to us. Phenomenology 'studies conscious experience as experienced from the subjective or first-person point of view' (Woodruff Smith 2018: online). Arendt does, however, overturn some of the thinking of her phenomenological predecessors such as Husserl and Heidegger. As explored earlier in this chapter, in Husserl's thinking the intersubjective experience leads to an understanding of others which emphasises what we have in common. In contrast, Arendt's understanding of intersubjectivity, as noted earlier, is based on the *uniqueness* of every subject and emphasises the value of 'bearing with' the 'strangeness' of the other. For Heidegger the authentic self is revealed more fully in separation from the world and its distractions, whilst for Arendt it is through being *with the other in plurality* that the self emerges in an intersubjective event (Parekh 2007: 70, Loidolt 2016: 46).

   For Arendt, then, with the roots of her thinking located in the phenomenological tradition, albeit with significant differences from her predecessors, approaching plurality is about a person going out into the world and experiencing its phenomena *as an event from within a first-person perspective*. Such an event engages with others in their *uniqueness* rather than conceiving 'the multiplicity of humans "from the outside" in their quantity or in their qualities and properties from a third-person perspective' (Loidolt 2016: 44). Plurality is not a static or substantial concept. Rather it is a plurality of unique '*who's*' – a plurality of first-person perspectives with the capacity in an encounter with an other to express their uniqueness, and where each is open to the expression of the other. Thus, for Arendt, plurality is *not* merely to do with quantitative multiplicity of people who are 'just there' or ready to hand (*vorhanden*), 'nor a quantitative or qualitative differentiation within a multiplicity, like unique genetic codes, different socialization processes, or multiculturally understood diversity' (Loidolt 2016: 44).

   It is the emphasis on a first-person perspective that leads Arendt (1974 [1958]: 176) to distinguish between *otherness* (*alteritas*), which arises because 'we are unable to say what anything is without distinguishing it from something else'; *distinctness*, which is 'the expression of the variations in organic life' and *uniqueness*, which is the capacity to 'express this distinctness and distinguish [one]self . . . to communicate [one]self and not merely something'. As Loidolt (2016: 44) points out:

> Whilst 'otherness' is an abstract universal property of every being (the scholastic *alteritas*) and 'distinctness' is an unconscious variation of life in living beings, 'uniqueness' implies living and self-aware beings who are able to express their stance. The latter is therefore a concept that involves an articulate first-person perspective, as well as other first-person perspectives that are receptive of the stance expressed.

The appearance of *who* a person is, then, needs to be actualised through the process of an *encounter* or event in which a subject can express their uniqueness to an other who is open to its expression and where one is also open to the expression of the other.

The emergence of who one is in an event also has implications for how one approaches the issue of human subjectivity. As Loidolt (2016: 49) explains, 'The mode in which the "who"' shows itself – and at the same time eludes the fixation of the "what" – is that of acting and speaking. It is intersubjective interaction'. Loidolt continues by highlighting how:

> The 'who' that appears in this interaction is not a representative, not a reflection of an already fully-fledged substantial 'inner self'. . . . Thus, who one is only develops in actualisation with others.

*Subjectivity **enacts** itself in the world through an intersubjective process*, creating, as I previously cited in Chapter 5, what Honig (1995: 146) calls:

> an agonistic disruption of the ordinary sequence of things that makes way for novelty and distinction, a site of resistance of the irresistible, a challenge to the normalising rules that seek to constitute, govern and control various behaviours.

Such *subjectivity is immanent*. It emerges in and through encounters in the 'web of relations' in the world: a web of relations also highlighted by feminist thinkers such as Tronto and Fisher (1991) and Haraway (2008, 2015, 2016) and many Indigenous thinkers[3] (as explored in Chapters 2 and 3). Buber (1958: 24–25) also emphasises immanent subjectivity emerging in and through embodied encounters. As cited in Chapter 3, he comments:

> The primary word *I-Thou* can be spoken only with the whole being. Concentration and fusion into the whole being can never take place through my agency, nor can it ever take place without me. I become through my relation to the *Thou*; as I become *I*, I say *Thou*.
>
> All real living is meeting.

Subjectivity which emerges immanently allows for the possibility of the emergence of a self which exceeds existing ways of being in the world and thus creates a 'surplus'. There is something more than was in the world before. This surplus is both inserted into the existing world and also rearranges that world – a *dissensus* – 'a dispute over what is given and about the frame within which we see something as given' (Rancière 2010: 69). The *dissensus* created by the immanent subject/self is thus an assertion of radical democracy since it is both an expression of freedom to act and also an unforeseen reconfiguration of the way the world is, and valid ways to be a subject within it.

Keller (1986) explores the notion of an immanent unique subject from a feminist perspective. She argues that immanence has been conflated in some feminist writing with the 'solubility' of the self: a self dissolved in the presence of the other, and thus seen as something to be resisted. She argues, however, for a positive understanding of immanence in which 'immanence is the way relations are part of who I am' (18), and where 'stagnation results from failing to "do a new thing" with and within the field of relations' (18). For Keller, an immanent subject evolves in an encounter, '*never outgrowing its open malleable ego boundaries*' (134). She proposes that an immanent subject can and should resist 'the predetermination of the subject as autonomous' (9) since focus on autonomy can, paradoxically, be a barrier to freely choosing what it is to be a human subject. Autonomy's sense of separation even separates the (hu)man from himself, a situation Keller calls the *separative* self, commenting:

> The separative self is identifiable historically, but neither essentially nor necessarily, with males and the masculine. Its sense of itself as separate, as over against the world, the Other, and even its own body, endows it with its identity. It is *this* not *that*.
>
> (8)

hooks (1994: 19) also highlights this loss of choice of ways to be a subject. She argues that patriarchy, with its emphasis on a rational, autonomous self, is damaging for men as well as women. She describes how, as children, both she and her brothers understood that 'we could not be and act the way we wanted, doing what we felt like': a situation which hindered the emergence of other possible ways of being and acting for them all. Bergson, too, challenges stable understandings of the subject since, 'it presumes the separation or discontinuity of the subject from the range [of possibilities] available' (Grosz 2011: 63). Through her nomadic theory (introduced in Chapter 4), Braidotti provides a posthuman perspective on immanent subjectivity in that her thinking includes the possibility of Keller's 'open malleable ego boundary' not only between humans but also between the human and the other-than-human, an intersubjectivity I explore in more detail in Chapter 8.

The ideas of Loidolt, Keller, hooks and Braidotti, as well as other feminist perspectives (for example see Plumwood 1993, 1995, Grosz 2011), and the Indigenous thinking explored in Chapters 2 and 3, help us to begin to play with and imagine that an abundance of unforeseen ways of being a human subject can emerge in and through intersubjective encounters. Engaging with these possibilities opens new creative spaces and subjectivities that valorise, amongst other possibilities, both rationality and emotion, mind and body, thought and matter, separation and connection, the human and the wider natural world of which humans are a part.

Problematising 'conditions of plurality'

Thus far I have argued that first-person intersubjective encounters can open up under conditions of plurality – that is to say, in spaces in which one can express one's uniqueness to others who are open to this expression and where one is also open and responsive to the expression of unique others. However, encouraging conditions where plurality is possible is not unproblematic. The thinker Bilgrami (2014) highlights how being open to 'unique others', to other ways of being in the world, should not be assumed as a 'given' possibility, especially when one is normalised in one's own particular ways of thinking. He argues that this situation is prevalent particularly in Western framings, which are often so dominant that one is unaware of both the existence of other ways and also the power dynamics which dominant ontologies and epistemologies operate to marginalise, silence and denigrate these other possibilities. This closes down openness to other potential subjectivities informed by ontologies and epistemologies of the Global South. He draws on the ideas and practices of Gandhi to explore an orientation towards the world informed by '*living an unalienated life*': a life informed by communality where liberty to achieve short term gain is not given centre stage and where one is able to recognise and respond to the normative demands placed upon us by the world – and all that is 'lively' within it.[4] He comments that many with 'worldviews' from the Global South note the contradictory and damaging thinking which dominant Western thinking places centre stage. These ideas are damaging both to the self and the world since they alienate the subject from the world. As an example of such Western contradictory thinking, Bilgrami cites the impossibility of achieving both liberty (emphasising autonomy) to maximise opportunities for oneself and also equality with its emphasis on equal shares and opportunities for all. Just as damaging is the Western assumption that dominant Western (Eurocentric) ways will, as a matter of course, be adopted by the rest of the world as they represent 'progress'. Such thinking needs to be challenged if conditions of plurality – an openness to the expression of the other – are to become a possibility. Whilst this is difficult, a first step towards being open to the expression of the unique other is to become aware of one's own assumptions, to do one's own 'homework[5]', to use Spivak's (1993) phrase, and also to develop one's capacity *to listen attentively* to the other. Such activities are explored in the next section, which considers ways to encourage first-person intersubjective encounters in educational settings.

First person intersubjective encounters and education

In this chapter I have argued that first-person encounters open up the possibility of spaces of appearance, in and through which who one is as a unique subject can emerge. The self that emerges in such a process is formed *intersubjectively* in that event. It is *not* that some kind of previously formed 'inner self' is revealed in such encounters. Rather, the other forms/becomes part of who I am and become. Empowering teachers and students to facilitate (whilst recognising they can never *produce*) 'first-person' intersubjective encounters opens up the possibility that unique subjects who emerge in such processes can insert themselves into the web of relations in the world and, to cite Arendt (1974 [1958]: 183–184):

> start a new process which eventually emerges as the unique life story of the newcomer, affecting uniquely the life stories of all those with whom he comes into contact.

However, facilitating intersubjective encounters in educational settings is very challenging. The temptation is to fill the space, to (pre)define the subject, to close down opportunities for natality. However, intersubjective encounters in educational settings *are* possible, as I now explore, drawing on theoretical and practical concepts and examples.

'Homeworking' and 'listening first'

As highlighted earlier, if education is to be a place where one is open to the expression of the other in first-person encounters, Spivak's (1993) 'homeworking' and the 'walking with' and 'learning from' highlighted by Sundberg (see Chapter 2) are important first steps. Blenkinsop et al. (2017)[6] recommend Memmi's (1991 [1957]) ground-breaking thinking – 'listening first' – as an approach which can help to encourage these steps. 'Listening first' requires developing 'our capacity *to deconstruct those elements interfering with our ability to listen*, while at the same time *cultivating actual listening practices and modes of engagement with other[s]'* (Blenkinsop et al. 2017: 353, emphasis added). Blenkinsop et al. call this the 'deferential practice of "shutting up"' (351): an acceptance that the one who is dominant in speaking has 'nothing to say and everything to learn' (361). This stands in contrast to the more dominant Western approach of trying to 'solve' the perceived problems of the other without opening a space for listening, responding to and giving the lead to marginalised others for finding ways forward.

Education as Skholé

As I discussed in Chapters 4 and 5, Education as *skholé*, can contribute to 'listening first' if the suspension proposed by Masschelein and Simons (2013) is extended to include suspension of the constraints placed on some students by dominant Western understandings which exclude other possible framings – other ontologies and epistemologies. Western teachers need to become aware of and challenge their own habitual thinking as part of this process. Such education as *skholé* can be a place to listen, a place to play with other possibilities – be they radically new or, to use MacFarlane's (2019) phrase, 'new-old' – ideas which have existed often for millennia but been ignored or denigrated. Received ideas and limitations can be questioned, challenged – made 'profane', to use Masschelein and Simon's (2013) term. 'Attentiveness' to the 'other' on this occasion can open up possibilities for new subjectivities to emerge in and through intersubjective processes. It is also important that when encounters do occur they are not closed down by the teacher eager to intervene or move on to the 'next thing'.

Memmi was writing in the context of the post-colonial situation in North Africa, but his ideas can fruitfully inform approaches to other marginalised groups, as teachers in a full-time government-funded education programme in the South West of England discovered. Working with a group of 'disaffected young people' aged between 16 and 18 identified by the UK Government as 'not in education, employment or training' teachers decided on a range of activities. These involved students planning and delivering events for various local groups in need of support or friendship. They felt this would engage students, develop literacy and numeracy (one of the key target areas for the UK government) and help them to be more outgoing and responsive to others, but they were disappointed by the students' lack of enthusiasm and told them so. When they started to listen to the students, however, to encounter them as unique individuals and be responsive to them, they discovered that each one was a young carer supporting a parent, sibling or grandparent, often in very demanding circumstances. These first-person intersubjective encounters profoundly affected the teachers and 'start[ed] a new process . . . affecting uniquely the life stories' (Arendt 1974 [1958]: 183) of all involved. In consultation with the young people, the activities were changed to ones in which students would plan, cost-out, then participate in a special trip such as to a bowling alley. These provided the young carers with opportunities to develop skills such as researching and budgeting and also a break from their responsibilities: a treat, a chance to share happy moments with others who understood the responsibilities each undertook.

Arendtian 'visiting'

Arendt's idea of 'visiting', introduced in Chapter 5, can also encourage a process of intersubjective first-person encounters. It can allow engagement with others in ways which try to minimise erasing their uniqueness. 'Visiting' does not reduce the other to the idea one holds of them in order to *presume* one can see through their eyes. Instead visiting opens up a space between the self and a unique other and then engages with the unique other in order to bridge that space, allowing new understandings and subjectivities to emerge in and through these intersubjective processes. The following case study provides an example of such visiting. This, and other teacher and student-devised activities, can be experimented with in education as starting points for encouraging the possibilities of intersubjective encounters.

Case study – The Challenge

The Challenge was a UK based national organisation which received government funding to work with young people through a range of programmes which have included the National Citizen Service (NCS), apprenticeship schemes and activities to encourage intergenerational interconnection. The organisation's vision was to encourage 'a more integrated society where there is understanding and appreciation of each other's differences' (The Challenge: online). Their mission was to 'design and deliver programmes that bring different people together to develop their confidence and skills in understanding and connecting with others'. These are broad aims which do not define success by certain outcome criteria such as increased employment rates, higher educational achievement, etc. Instead they allow for the possibility of emergence of different ways that participants can develop who they are as unique beings in 'the web of human relationships' (Arendt 1974 [1958]: 184). The Challenge did not, however, assume that achieving these aims was easy and recognised the need to include activities to facilitate openness to the other rather than subsuming the other into one's own ways of thinking and being. For example, the following activity was used in the first week of its NCS programme. Called 'Trust on a Scale', the activity asked participants on the programme to work in small groups to collectively place certain people listed on activity cards (e.g. parents, siblings, the police, teachers, doctors) on a scale, drawn on a large sheet of paper. The scale ranged from 'trust completely' at one end to 'don't trust at all' at the other. Even though each group only had one card per 'category' (e.g. parents, teachers) to add to the scale, the aim of the activity was not to come up with an 'answer' on 'appropriate trust levels'. Instead the activity was designed to open a space for awareness and openness to one's own and others' experiences around trust. This revealed how others' viewpoints and ideas can be surprisingly different from one's own (often unexamined) experiences and encouraged participants to develop their skills and confidence to understand and connect with others; reflect, understand and 'be' in new ways and explore what opportunities emerge in this process for their own and others' futures.

This activity can be emotive and raise difficult issues, for example, around trust levels in the police. Educators need to approach activities of this kind with care. But nonetheless the activity provides a helpful example of the possibility of intersubjective encounters in education and a concern for, and appreciation of, what can emerge through such encounters. Mouffe's (2005, 2007, 2014) conception of agonistic pluralism, introduced in Chapter 5, can also contribute here. It allows one to approach the other in an encounter not as an enemy but as an adversary with whom one can argue without trying to erase them and their different points of view. It allows conflict into encounters, accepts it as part of life in plurality, keeps it in play as something to explore together rather than attempting (unsuccessfully) to erase it. If agonism is not allowed in in this way, it will not disappear but instead remain unacknowledged at the margins: a trace of ways forward not selected – a situation which can develop into *antagonism*.

The encounters examined in this case study are between young people from different backgrounds – cultural, religious and economic. Intergenerational encounters can also be encouraged[7] with space and time to engage with each other as unique subjects, opening opportunities for new understandings and ways of being. Such encounters can challenge stereotypes (and sometimes fear) each may hold and start something new which changes all involved.

Dissensus

Rancière's conception of *dissensus*[8] provides a further way to encourage possibilities for intersubjective encounters in educational settings. For Rancière (2010: 69), a political subject has 'a capacity for staging scenes of *dissensus*' and for opening up the 'space of a test of verification' in the existing structuring of the sensible[9] world: ideas which I feel resonate with this case study on the charity Jamie's Farm.

Case study: Jamie's farm

Jamie's Farm is a UK registered charity whose vision is that through participation in their programme 'vulnerable young people will be better equipped to thrive during secondary school and beyond'. The 'initial spark' for the charity occurred when the founder Jamie Feilden started to teach in a challenging comprehensive school in the London Borough of Croydon, UK. Shocked by 'the battleground that the school had become' (Jamie's Farm 2020a: online), Jamie made an insertion which challenged the existing structures and 'arrangement of the sensible' in the school. He decided to bring some lambs to the school from his farm in Wiltshire and set up sheep-pens for them in the playground, inviting students to get involved in their care. This was an insertion which was both 'disruptive and inaugurative' (Dikeç 2013) to the situation he found himself in and one which disputed ways that behavioural issues are more frequently dealt with in education. He observed that it was often the students who had the greatest behavioural and emotional challenges who responded to and benefited the most from his invitation.

From these beginnings grew the UK based Jamie's Farm charity consisting of several rural working farms and an urban project at Waterloo, London. These farms are run on the principles of Farming+Family+Therapy and the programme is built around five-day residential stays with follow-up involvement with students and their schools. Fundamental to the approach is the assumption that the participants – vulnerable young people often at risk of exclusion from school – are able to do the various tasks involved in running the farm: an 'assumption of equality of intelligence' (Rancière 1991) and of practical and emotional capability. For example, children from inner city areas are actively involved in delivering lambs, with the farm assistant firm in his belief they could do this despite it being very different from anything they had experienced before. Other activities involve teamwork with staff and students, such as cleaning out and feeding farm animals, preparing and clearing up after healthy meals and hiking. Individual therapies, for example, with horses and a trained therapist, are also included.[10] Through such encounters with other humans and also the wider natural world, participants find a new sense of self emerging: one that both surprises and pleases them. They then carry this new self forward into their future ways of being and acting when they return to school. The unique subjects who appear thus insert themselves into the existing arrangement of the world, creating something new, something unforeseen. Whilst not wanting to limit their work to outcomes and statistics, the charity has recognised these are necessary for various reasons, such as attracting funding. In their impact studies they have found that the majority of students have improved self-esteem and mental wellbeing scores, and school staff who have been to the farm feel more able to work effectively with disadvantaged students. According to their most recent impact figures (relating to academic year 2017/18), six months after the programme, 53% of participants displayed improved behaviours; 58% of those at risk of permanent exclusion were no longer at risk, 66% displayed improved engagement and 80% of those referred over concerns around attainment of expected school level showed an increase in this (Jamie's Farm 2020b: online).

Jamie's Farm can also be understood as education as *skholé*. It offers a space apart, a suspension of the usual world and expectations placed on students. For example, students do not have access to sugary food or their phones or other computer devices during their stay to avoid the distractions and pressures these create. They are at first very dubious as to whether they can survive five days without these but usually soon come to terms with the situation and even comment on enjoying it as liberating. At Jamie's Farm there is always 'something on the table', to use Masschelein and Simon's (2013) term, but at the farm, this can be many kinds of things in all kinds of places! There is always something to encounter, to pay attention to, to respond to in new and unexpected ways.

Not everyone has lambs to keep in the school playground (although some do and this might not be particularly surprising, in certain places and contexts). However, in exploring this case study teachers and students can consider opportunities that they can access for encountering, and *dissensus* in their own situations. This sharing of ideas can itself be part of the process and potentiality of encountering.

It is important to recognise that not all encounters are cordial, and what emerges through encountering can be problematic. As I discussed in The Challenge case study, Mouffe (2007) reminds us of the need to accept agonistic pluralism as part of life and to allow it into spaces of encountering. Sensitivity to the ethical issues which agonistic pluralism can raise is important and it is to this I now turn.

Ethics and intersubjective encounters

In Chapter 6, and at the start of this chapter, I have highlighted how Arendt understands ethics not as a theory but as an event, an encounter, because it 'responds to specific difficulties in a particular context and is based on actual mutual confidence in the faithfulness and determination of others' (La Caze 2014: 217). First-person intersubjective encounters taking place in the 'web of relations' in the world, have the potential to be places to enact a horizontal ethic between people arising in and through processes of forgiveness and mutual promising rather than responding to situations with vengeance. Such a web is also foregrounded by feminist thinkers such as Tronto and Fisher (1991) and Haraway (2008, 2015, 2016) and many Indigenous thinkers such as Little Bear (2000, 2016) and Bawaka Country *et al.* (2016).

Whilst it can be difficult to have confidence in 'the faithfulness and determination of others' to forgive and make mutual promises, responding to situations in this way can be an unexpected expression of freedom and thus a democratic move. Breaking the cycle of vengeance releases one from the past and opens possibilities for new futures, new ways to be together. Arendt emphasises that the power potential to break the cycle of vengeance lies *between*, and can be *generated by*, people speaking and acting together: a power which mutual promising can then hold open. Intersubjective first-person encounters in which unique beings can speak and act with others, forgive and make mutual promises in response to specific ethical concerns and situations can contribute to developing the power to act further, to challenge damaging approaches and to suggest new, more sustainable ways to be in the world we share.

Whilst the focus of this chapter has been intersubjective encounters with other humans and the possibility of encouraging such encounters in education, the next step, and the focus of the next chapter, is to explore the possibility of intersubjective encounters within the wider natural world. Moreover, as the case study on Jamie's Farm indicates, encounters between humans and between humans and the wider natural world are entwined – an issue which the next chapter also examines.

Notes

1 Under conditions of plurality – when there is opportunity to speak and act with others where each is open to the stance the other expresses.
2 Solipsism – theory holding that the self can know nothing but its own modifications and that the self is the only existent thing (Merriam Webster: online).

3 For example, see Little Bear 2016, De Line 2016, KARI-OCA 2 Declaration 2012, Bawaka Country *et al.*

4 The idea of a lively world is explored in Chapter 8, where I examine the enchantment and vibrancy of the wider natural world.

5 Homeworking – 'locating our body-knowledge in relation to the existing paths we know and walk' (Spivak 1993: 39).

6 In this article, Blenkinsop *et al.* draw on Memmi's ideas to challenge human colonisation of 'nature'. The possibility of intersubjective encounters with the wider natural world is the focus of Chapter 8, where there are further discussions of the article.

7 The Challenge supported intergenerational meeting as part of its programmes. Encouraging intergenerational encounters is also a regular feature of programmes for disaffected young people at a college of further education in the South West of England.

8 As introduced in Chapter 5, *dissensus*, for Rancière (2010a: 69) is 'not a conflict of interests, opinion, or value; it is a division inserted in "common sense": a dispute over what is given and about the frame within which we see something as given'.

9 I.e. the world one senses.

10 I explore the possibility and impact of intersubjective encounters with the wider natural world experienced by the participants in the next chapter.

References

Arendt, H. (1974 [1958]) *The human condition*. Chicago, IL: University of Chicago Press.

Arendt, H. (1994 [1954]) Understanding and politics. In *Essays in understanding*. New York: Schocken Books.

Bawaka Country, Wright, S., Suchet-Pearson, S., Lloyd, K., Burarrwanga, L., Ganambarr, R., Ganambarr-Stubbs, M., Ganambarr, B., Maymuru, D. and Sweeney, J. (2016) Co-becoming Bawaka: Towards a relational understanding of place/space. *Progress in Human Geography*, 40(4): 455–475.

Biesta, G. (1994) Education as practical intersubjectivity: Towards a critical pragmatic understanding of education. *Educational Theory*, 44(8): 299–313.

Biesta, G. (2006) *Beyond learning: Democratic education for a human future*. Boulder, CO: Paradigm Publishers.

Bilgrami, A. (2014) *Secularism, identity and enchantment*. Cambridge, MA: Harvard.

Blenkinsop, S., Affifi, R. Piersol, L. and De Danann Sitka-Sage, M. (2017) Shut up and listen: Implications and possibilities of Albert Memmi's characteristics of colonisation upon the 'natural world'. *Studies in Philosophy and Education*, 36(3): 349–365.

Braidotti, R. (2011a) *Nomadic theory: The portable Rosi Braidotti*. New York: Columbia University Press.

Braidotti, R. (2011b) *Nomadic subjects: Embodiment and sexual difference in contemporary feminist theory*, 2nd Edition. New York: Columbia University Press.

Buber, M. (1958 [1923]) *I and thou*. Kaufmann, W. (tr.). New York: Scribner.

De Line, S. (2016) All my/our relations: Can posthumanism be decolonised? *Open! Platform for Art, Culture & the Public Domain*. Available at: www.onlineopen.org/all-my-our-relations [Accessed 23.1.2018].

Dikeç, M. (2013) Beginners and equals: Political subjectivity in Arendt and Rancière. *Transactions of the Institute of British Geographers*. Royal Geographical Society (with the Institute of British Geographers), 38(1): 78–90.

Grosz, E. (2011) *Becoming undone: Darwinian reflections on life, politics and art*. Durham, NC: Duke University Press.

Habermas, J. (1984) *The theory of communicative action*, Vol. 1. Boston, MA: Beacon Press.

Haraway, D. (2008) *When species meet*. Minneapolis and London: University of Minnesota Press.

Haraway, D. (2015) Anthropocene, Capitalocene, Plantationocene, Chthulucene: Making kin. *Environmental Humanities*, 6: 159–165.

Haraway, D. (2016) *Staying with the trouble: Making kin in the Chthulucene*. Durham, NC: Duke University Press.

Honig, B. (1995) Toward an agonistic feminism: Hannah Arendt and the politics of identity. In B. Honig (ed.) *Feminist interpretations of Hannah Arendt* (pp. 135–166). Philadelphia: Pennsylvania State University Press.

hooks, b. (1994) *Teaching to transgress: Education as the practice of freedom*. New York: Routledge.

Jamie's Farm (2020a) *About us: Our journey*. Available at: https://jamiesfarm.org.uk/about-us/our-journey/ [Accessed 18.3.2020].

Jamie's Farm (2020b) *About us: Our impact*. Available at: https://jamiesfarm.org.uk/about-us/our-impact/ [Accessed 18.3.2020].

KARI-OCA 2 Declaration (2012) *KARI OCA 2 declaration: Indigenous peoples global conference on RIO + 20 and mother earth*. Accepted by Acclamation, Kari-Oka Village, at Sacred Kari-Oka Púku, Rio de Janeiro, Brazil, 17th June. Available at: https://wrm.org.uy/other-relevant-information/kari-oca-2-declaration-indigenous-peoples-global-conference-on-rio-20-and-mother-earth/ [Accessed 17.12.2019].

Keller, C. (1986) *From a broken web: Separation, sexism and self*. Boston, MA: Beacon Press.

La Caze, M. (2014) Promising and forgiveness. In P. Hayden (ed.) *Hannah Arendt: Key concepts*. Durham, NC: Acumen, pp. 209–221.

Little Bear, L. (2000) Jagged world views colliding. In M. Batisse (ed.) *Reclaiming indigenous voice and vision*. Vancouver: University of British Columbia Press. Available at: http://www.learnalberta.ca/content/aswt/worldviews/documents/jagged_worldviews_colliding.pdf [Accessed 14.10.2019].

Little Bear, L. (2016) *Blackfoot metaphysics 'waiting in the wings'*. Congress of the Humanities and Social Sciences Big Thinking Lecture, 1st June. Available at: www.youtube.com/watch?v=o_txPA8CiA4 [Accessed 18.11.2018].

Loidolt, S. (2016) Hannah Arendt's conception of actualised plurality. In D. Moran and T. Szanto (eds.) *Phenomenology of sociality: Discovering the 'we'*. London: Routledge, pp. 42–55.

Macfarlane, R. (2019) Should this tree have the same rights as you? *The Guardian*, 2nd November. Available at: www.theguardian.com/books/2019/nov/02/trees-have-rights-too-robert-macfarlane-on-the-new-laws-of-nature [Accessed 7.11.2019].

Masschelein, J. and Simons, M. (2013) *In defence of schools: A public issue*. Leuven: Education, Culture and Society Publishers. Available at: http://ppw.kuleuven.be/ecs/les/in-defence-of-the-school/masschelein-maarten-simons-in-defence-of-the.html [Accessed 3.2.2016].

Memmi, A. (1991 [1957]) *The coloniser and the colonised*. Introduction by Jean-Paul Sartre, afterword by Susan Gilson. Greenfeld, H. (tr.). Boston, MA: Beacon Press.

Merriam Webster (online) *Solipsism*. Available at: www.merriamwebster.com/dictionary/solipsism [Accessed 18.7.2020].

Mouffe, C. (2005) *The democratic paradox*. New York: Verso.

Mouffe, C. (2007) Artistic activism and agonistic spaces. *Art and Research: A Journal of Ideas, Contexts and Methods*, 1(2). Available at: www.artandresearch.org.uk/v1n2/mouffe.html [Accessed 9.2.2015].

Mouffe, C. (2014) Democracy, human rights and cosmopolitanism: An agonistic approach. In C. Douzinas and C. Gearty (eds.) *The meaning of rights: The philosophy and social theory of human rights*. Cambridge: Cambridge University Press, pp. 181–192.

Parekh, S. (2007) *Hannah Arendt and the challenge of modernity*. London: Routledge.

Plumwood, V. (1993) *Feminism and the mastery of nature*. London: Routledge.

Plumwood, V. (1995) Human vulnerability and the experience of being prey. *Quadrant*, 29(3): 29–34.

Rancière, J. (1991) *The ignorant schoolmaster: Five lessons in intellectual emancipation*. Ross, K. (tr.). Redwood City, CA: Stanford University Press.

Rancière, J. (2010) Who is the subject of the rights of man? In J. Rancière (ed.) *Dissensus: On politics and aesthetics*. London: Continuum, pp. 70–83.

Spivak, G. (1993) *The post-colonial critic: Interviews, strategies, dialogues*. New York: Routledge.

Sterling, S. (2010b) Living in the earth: Towards an education for our time. *Journal of Education for Sustainable Development*, 4(2): 213–218. doi: 10.1177/097340821000400208

The Challenge (online) *What we do*. Available at: https://the-challenge.org/what-we-do/ [Accessed 25.10.2019].

Topolski, A. (2015) *Arendt, Levinas and a politics of relationality* (Reframing the boundaries: Thinking the political). London: Rowman & Littlefield International.

Tronto, J.C. and Fisher, B. (1991) Toward a feminist theory of caring. In E. Abel and M. Nelson (eds.) *Circles of care*. Albany, NY: SUNY Press, pp. 36–54.

Woodruff Smith, D. (2018) Phenomenology. In E.N. Zalta (ed.) *Stamford encyclopaedia of philosophy*, Summer 2018 Edition. Available at: https://plato.stanford.edu/entries/phenomenology/ [Accessed 29.1.2020].

8 Subjectivity and intersubjective encounters within the wider natural world

Introduction

In this chapter I explore the possibility of intersubjective *first-being* encounters between humans and the *wider natural world* and how such encounters could be encouraged, including in educational settings. The emphasis is not on creating a totalising theory of subjectivity or intersubjectivity. Rather, the approach adopted focuses on specific 'first-being' *encounters* and the possibility of the emergence of radically new, unique subjectivities through these encounters. In this chapter I use the term 'first-being' to replace the grammatical term 'first-person' As in earlier chapters I use the term 'wider natural world' to indicate that humans are themselves part of this wider natural world.

The possibility of intersubjective first-being encounters between humans and the wider natural world is a departure from the work of Arendt who, writing from her own particular era and framing, limits her attention to the web of *human* relations and places humans as '*unlike*' other living or inanimate things' which 'merely exist' (Arendt 1974 [1958]: 199). This is a position that this chapter challenges. Yet I would like to think that Arendt would be open to the possibilities that such a move creates. It responds to her own invitation to recognise and value the importance of engaging with ideas which have been developed in previous eras without allowing them to become a 'chain which fetters us' (Arendt 2006 [1961]: 94). Questioning the authority of the human and the belief that humans have a unique capacity for subjectivity whilst other aspects of the world 'merely exist' enables a much broader understanding of subjectivity in 'others unlike us' (Pedersen 2010: 243). This has implications for educational approaches which seek to encourage the opening of spaces in and through which unique subjectivities can appear. A rethinking of dominant Western education understood as *a '*huma*nities subject'* is needed.

I acknowledge that the rethinking of Western boundary-keeping examined here is done from my position within it, accepting Bonnett's point (2000, 2002, 2004) that it is not possible to escape one's own worldview sufficiently to understand and represent *fully* ideas coming from different onto-epistemological starting points. It is for this reason that I mainly draw on Western theorists and practitioners, albeit they are ones actively critiquing Western framings of the world. These thinkers are engaging in what Spivak (1993) calls 'homeworking' – (examining one's own intellectual roots as a necessary first step before engaging with others) as well as engaging with ideas from other ontologies and epistemologies to develop their own thinking. I also draw in thinkers who are writing from the borders of Western philosophy, for example Buber, and Indigenous knowledges and ways of being, particularly the emphasis in many of these on the importance of specific encounters and attentive and caring inter and intra-connections in such encounters. However, following Todd (2016), and recognising the point made by Bonnett, I would encourage a direct reading of Indigenous thinkers to expand understanding of and engagement with these ideas. As in Chapter 7, I also include some practical ideas and case studies deliberately drawn from initiatives with which I have had *my own first person/first-being encounters*. These are initiatives working at the 'margins' of mainstream UK education: margins where I have spent my own teaching life.[1] The case studies provide opportunities for reflection and discussion as well as starting points for consideration of teachers' and students' own first-being encounters in their own settings.

This chapter examines in more detail recent Western debates (introduced in Chapter 4) on subjectivity in the wider natural world. I then consider the possibility of intersubjective first-being encounters between humans and the wider natural world and the potential such encounters can open up. Drawing on a range of starting points such as touch, entanglement, listening and attentiveness, slow education, the value of pausing, 'enchantment' and 'wild pedagogies', the final part of the chapter explores ways that intersubjective first-being encounters can be encouraged, although never produced or guaranteed in educational settings.

Subjectivity and the wider natural world

'Subjectivity' in Western thinking refers to a sense of self, a sense of who one is and how one acts in the world. It includes having capacity to reflect on selfhood and have higher order thoughts. The possibility that animals, plants and other participants in the natural world, such as water and rocks, have capacity for subjectivity is a current topic of debate in academic and other literature. A key point for me in this debate is that *acknowledging subjectivity beyond the human does not require that it 'be similar in all respects to one's own'* (Lyvers 1999: 5). As Lyvers[2] points out, assuming that only humans are capable of subjectivity, as is the case in the dominant Western philosophical tradition, is a stance which wishes to place humans at the centre of the universe, above other elements of the natural world. He argues that this is akin to the old notion that the earth was at the centre of the solar system – an idea which seems absurd today. In Chapters 2 and 3 I argued that it is one's worldview that form the basis for one's ways of being in (ontologies) and knowing (epistemologies) the world. I recognise this is not an easy argument for some to accept. Some find even the suggestion of alternative worldviews in which, for example, all is animate with potential for subjectivity, tremendously problematic, or worse, something to be dismissed or derided. For others, however, either intuitively, or through openness to different ideas and ways of being, the possibility and validity of other worldviews is part of life and it is such possibilities and their implications that this chapter explores.

In his engagement with the issue of subjectivity, Massumi (2014) highlights our present separation from the rest of the animal spectrum, our vanity regarding our assumed species identity and our unique ownership of language, thought and creativity. In contrast, he places humans on the animal spectrum, arguing that rather than denying our 'animality', engaging with this proposition can help inform human subjectivity. Massumi recognises that 'expressing the singular belonging of the human to the animal continuum has political implications, as do all questions of belonging' (3). He realises that proposing such an approach opens him to the accusation of anthropomorphism, but comments he would willingly risk such an accusation:

> in the interests of following the trail of the qualitative and subjective in animal life, and of creativity in nature, outside the halls of science, in the meanders of philosophy, with the goal of envisioning a different politics, one that is not a human politics of the animal but an integrally animal politics, freed from the traditional paradigms of the nasty state of nature and the accompanying presupposition about instinct permeating so many facets of modern thought.
>
> (2014: 2)

Derrida (2002, 2008) identifies that the (Western) categorisation of humans and animals creates a false binary. He argues that humans have given themselves the authority to name all others under a single word – 'animal'. This Western human-animal distinction groups all animals together as though they share a single way of being in and experiencing the world; they only react to external stimulus and have no access to their own subjectivity. Moreover, '[t]he very history of who we think we are as humans is tied up in distinguishing ourselves from this "other"', which we have dominated/exploited, 'claiming subjectivity as our exclusive property' (Derrida 2008: 23). Furthermore, although the Western human-animal binary and the human domination/exploitation of animals which arises from it has not changed, what *has* changed 'is the unprecedented proportions' of this domination. Derrida also comments: 'Neither can one seriously deny the disavowal that this involves' (25) – a disavowal Derrida is not prepared to participate in for the 'putative well-being of man' (25). Moreover, as Weil (2008: 2) points out, Derrida is not afraid 'to use the words that others may have shied away from – holocaust, genocide – to describe in detail the kinds of violence done to animals through industrial farming or biological experimentation and manipulation'. Those who argue for ecocide[3] to be recognised internationally as an atrocity crime are also not afraid to use language in this way. They extend the consideration of the violence identified by Derrida as directed towards animals to all participants in the wider natural world. Such violence is, for example, noted in the United Nations Intergovernmental Science-Policy Platform (IPBES) *Biodiversity and Ecosystem Services Assessment* (2019: online) which states, 'Nature is declining globally at rates unprecedented in human history and the rate of species extinctions is accelerating'. Western (Eurocentric) categorising of nature as an object, a resource to be used, a lesser 'other', has significantly contributed to this rapid decline.

Klein (2015) and Klein and Lewis (2015) foreground the link between capitalism (with its roots in the Western Industrial Revolution) and this categorisation by humans of the 'non-human world' as a resource, an object owned by humans rather than as subjects with agency (i.e. having subjectivity). Klein and Lewis argue that this derives from a sense of separation from, and also a domination over, the wider natural world (a separation which Plumwood [2001] calls 'hyperseparation') founded in a worldview heralded by Western (Eurocentric) mechanistic philosophers such as Newton and Boyle. As discussed in Chapter 2, in this framing 'nature' is understood as 'brute and inert', governed by a mechanical or clockwork process with humans increasingly understood as masters of the mechanism. In the period between the development of this mechanical worldview (which broadly speaking occurred in the seventeenth century) and the Western Industrial Revolution (which broadly speaking began in the eighteenth century in Britain) this human sense of domination was held in check, to some extent, because humans still needed to bend to the vagaries and power of nature. For example, the wind was needed to power ships. However, the Western Industrial Revolution introduced technical developments which then enabled humans to exploit the wider natural world seemingly unchecked, an exploitation that spread around the world through colonialism and laid the foundations for capitalism.

The painter J M W Turner lived through and recorded this period of history in his painting. His work documents 'the "taming" of the sublime – the awe-inspiring dangerous-yet-beautiful power of nature – by the hand of Man' (Beavan 2016: online). This 'taming' can be seen in Turner's 1819 painting of Bell Rock Lighthouse (see Figure 8.1) which stands off the west coast of Scotland and still operates today. This painting was commissioned by the engineer Robert Stevenson to illustrate his publication documenting the design and construction of the lighthouse. Prior to this, shipwrecks were common at Bell Rock, and ships and their scheduling were dependent on the weather. The lighthouse was (and still is) a feat of engineering, able to withstand the power of seas which in stormy weather made it hard to navigate around the rocks as they are completely covered at high tide. In his painting Turner depicts ships sailing safely by the reef despite the raging seas: the 'nasty' state of nature – tamed and put to the service of man.

Klein (2015) highlights how it was this meeting of technology with the ideology of human domination over nature which enabled capitalism to flourish. However, recent events, such as the extensive fires in Australia, extreme

Figure 8.1 Bell Rock Lighthouse

[Medium: Watercolour, gouache and scraping out on paper] (1819), Joseph Mallord William Turner, The National Gallery of Scotland: Edinburgh

Source: [Reproduced with permission of National Galleries of Scotland. Purchased by Private Treaty Sale 1989 with the aid of funds from the National Heritage Memorial Fund and the Pilgrim Trust]

pollution in China leading to severe smog, the Covid-19 pandemic and developing understandings of climate change and its impact, highlight how humans are not 'in charge of nature'. This realisation is contributing to debates over limits of, and resistance, to capitalism, in both the Global North and the Global South. For example, in 2019, New Zealand's Prime Minister, Jacinda Ardern, set a 'well-being budget' based on citizens' health and life satisfaction to measure a country's progress rather than using wealth or economic growth calculations such as gross domestic product (GDP). She commented, 'GDP alone does not guarantee improvement to our living standards' and nor does it 'take into account who benefits and who is left out' (Samuel 2019). New Zealand is not the first country to consider such an approach. In the 1970s, the king of Bhutan used the term 'gross national happiness' to assert that 'Gross National Happiness (GNH) is more important than Gross Domestic Product' (Samuel 2019). This interest in GNH was developed and enshrined in Bhutan's constitution in 2008. Prempeh (2008) highlights anti-capitalism resistance in the African continent occurring at the same time as this stand in Bhutan against the dominance of GDP. Such resistance continues today across the Global South, as documented, for example, in the 2018 Special Issue of the *Agrarian South: Journal of Political Economy* and in the work and writing of Indian conservationist Prerna Singh Bindra (2019).

In their exploration of agency and subjectivity in the wider natural world, Walsh, Karsh, and Ansell (1994) – an interdisciplinary team consisting of a world studies academic, a philosophical theology academic and a forester – focus on trees. They too challenge capitalism's and Western/modernist philosophy's identification of nature as an object and resource for humans which 'supposes that nature is an It' (149). Instead they 'see all creation as a "Thou" awaiting subject-to-subject relatedness with humankind' (150). To develop their argument, Walsh *et al.* (1994) draw on Borgmann (1992: 117) who points out how:

> postmodern theorists have discredited ethnocentrism and logocentrism so zealously that they have failed to see their own anthropocentrism. Why reject *a priori* the very possibility that things may speak to us in their own right?

Borgmann (1992: 118–119) continues, 'Rivers are muted when they are dammed; prairies are silenced when they are stripped for coal; mountains become torpid when they are logged'. To respond to this situation Borgmann proposes a 'postmodern realism', which is 'not the naive and aggressive realism of modernity' but instead 'an attending to the eloquence of reality'. Without such attention, Walsh *et al.* (1994: 149) argue that 'there can be no response to the other's cry and no learning from the other's wisdom'.

Walsh *et al.* (1994) discuss Buber's reflection on his encounter with a tree which opened up 'an I-thou reciprocal relationship' (1958 [1923]: 7) rather than an 'I–it relationship' in which trees are interchangeable and 'just there' for human use. Walsh *et al.* comment that when *I and Thou* was first published, many raised within a Western Enlightenment rational autonomous worldview found Buber's reflections on his encounter with a tree very challenging, particularly the issue of reciprocity. They note how if we do:

> enter into *I-Thou* relations with trees, doing so would not be a romanticist projection of relatedness, but a conviction that in some important way trees reciprocate the relationship; that not only do we relate to trees, but they also relate to us.
>
> (150)

They continue, 'now the worldview that presupposed an objectified nature has run its disastrous course . . . we are open to a different way of relating, a different way of life, beyond the subject/object dualism, beyond the *I-it* relationship' (151). They argue we are now more open to ways of life which challenge 'the autonomous and imperial ego' and which 'strive anew for contact, for reciprocity – and that our striving must aim at what Buber (1958 [1923]: 79) called "tenderness"'. They argue, 'We want to *listen* to Buber again. Even more, we want to learn to *listen* to the trees' (151). This, however, is problematic since having struck trees 'deaf and dumb' (to use Borgmann's phrase) for so long, Western minds have lost the ability to listen in this way.[4]

Engaging with the wider natural world as responsive and reciprocal has generally been lost in Western framings, although it continues within many Indigenous knowledge systems. Walsh *et al.* provide examples of ways that responsiveness and reciprocity can be regained in Western framings. They discuss the work of biologist and forestry expert J.L. Shigo and his idea of *attentiveness;* environmentalist John Muir's emphasis on *listening informed by both botanical knowledge and empathy* and the attitudes towards plants of geneticist Barbara McClintock in her work on the identification of genetic transposition[5] in plants which allows an entity to be responsive to its environment.[6] When McClintock first published her results in the 1940s, they were not believed, but many years later, in an environment more open to such ideas, they became better understood. This led, in 1983, to her being awarded the Nobel Prize in Physiology or Medicine. Walsh *el al.* comment that McClintock's approach to her scientific work required an attitude of involvement and attentiveness to corn plants, a kindred subjectivity which we normally only associate with relationships with other humans. This attentiveness and kindred subjectivity is an important aspect of the embodied ways of being and co-becoming within an intra-connected world of many Indigenous people, for example in the lives of the Indigenous people in Bawaka Country, North East Arnhem Land, Australia whose ideas I introduced in Chapter 3 and discuss in more detail later in this chapter.

Even though some thinkers, such as Noddings (2013), argue that plants do not have the subjectivity and agency necessary for reciprocity (which for Noddings is an essential part of caring), Walsh *et al.* (1994) argue otherwise. They propose that such a denial focuses on rational decision making. However, if one takes a different starting point, one that recognises that humans are 'multidimensional creatures and our intellectual capacities are but one aspect in the exercise of our will' (160), new ways of understanding agency can emerge which also draw out such possibilities within the wider natural world. They argue that attentiveness to trees indicates that trees actively respond to their environment in ways which exceed mechanistic explanations and require a subject-to-subject ethical engagement. This responsiveness and the emotional and community aspects of how trees live is now becoming increasingly understood and spoken about. For example, Wohlleben's (2016) *Secret Lives of Trees* (discussed in Chapter 3) has become an international best-seller. This shift of attention and interest is also reflected in the British Broadcasting Corporation's (BBC) decision in 2019 to produce a five-part series, *The Green Planet*, focusing on the life of plants (rather than animals). Presented by David Attenborough and using a variety of new technologies, it will be, according to the

BBC, 'the first immersive portrayal of an unseen, inter-connected world, full of remarkable new behaviour, emotional stories and surprising heroes in the plant world' (BBC Media Centre 2019).

Thus far in this chapter, I have explored subjectivity beyond the human and ideas and experiences which argue that such subjectivity and reciprocity between humans and the wider natural world are possible and can be encouraged. This is particularly so if one accepts that whilst subjectivity in the-other-than human may be different from human subjectivity, it is nonetheless just as valid. This opens ways for encounters between humans and the wider natural world to be understood in subject-to-subject terms and the possibility that such encounters can be intersubjective processes in and through which new and unexpected subjectivities can emerge. This possibility needs further consideration, and it is to this I now turn.

Intersubjective encounters within the wider natural world

Whilst recognising the dangers of 'generalising' about Indigenous thinking, many Indigenous epistemologies/ontologies emphasise encounters and the dynamic relationships within the web of life that such encounters can encourage.[7] Encounters are specific and multisensory: the encounter is with this rock as a shelter for a plant, or next to a snake's nest, in this specific place, and the human sense of touch, taste or smell or sense of fear are an integral part of this experience (see discussion in Kuletz 1998, Deloria 1999, Watts 2013). The ideas developed in this chapter acknowledge and respect this thinking and way of being. Such encounters can open a space in and through which new subjectivities, new ways of being together in the world, can emerge. As discussed in Chapter 7, 'The mode in which the "who"' shows itself – and at the same time eludes the fixation of the "what" – is that of acting and speaking. It is intersubjective interaction' (Loidolt 2016: 49). In first-being encounters with(in) the wider natural world, 'acting' can include listening, attentiveness, bodily entanglement. Moreover, the 'who' that appears in and through such intersubjective encounters 'is not a representative, not a reflection of an already fully-fledged substantial "inner self"' (49). Rather, such new subjectivity *enacts* itself in the world through the intersubjective process. It generates an 'agonistic disruption' (Honig 1995: 146) in the existing world, and a potential for *dissensus* – 'a dispute over what is given and about the frame within which we see something as given' (Rancière 2010: 69), which then changes that given framing.

Ground (2013) draws on Wittgenstein's conception of 'expressive enactment' to provide an approach to such first-being intersubjective encounters. Expressive enactments take place as first to second person events (I and Thou). The 'mutual reciprocity' (29) inherent in expressive enactments precedes and resists analysis from a third person perspective, especially analysis and models which rely on the use of a spoken language 'to act as a proof of other minds'

(29). The important consideration is to *be on the inside* of such exchanges which exist in the *relationship between* those engaged in the enactment rather than to attempt to explore from the outside (particularly in words), in independent mental constructs in the 'mind' of either party. The authors of the paper Bawaka Country *et al.* (2016) attempt to capture this sense of exploring from within an experience when engaging with *gurrutu* – a complex kinship system that extends beyond the notions of human kin to kinship with the more-than-human world. As I discussed in Chapter 3, they gathered *ganguri* (yams) and told the story of this experience: an embodied process rather than engagement with the concept in abstract academic language. The land is included as a co-author of the paper.

The *Democracy's Non-Human and Non-Living "Others" International Workshop, New York May 2nd, 2014* provided a forum for researchers to discuss engaging in different projects to listen to the other-than-human and to draw these 'voices' (expressed in a variety of ways other than through human language) into democratic practices.[8] The projects, as well as ideas from Chapters 2 and 3, provide interesting starting points for encouraging engagement in expressive enactments which have also the potential to be sites of ethical reciprocity. I highlight three here.

Bastian (2014) discussed a project funded by the Arts and Humanities Research Council (AHRC) via the Connected Communities Project, UK. Bastian's project considered ways to hear the dogs, bees, trees and water in the design process of participative research. This involved a range of embodied encounters as a starting point and required suspension of some 'habits of mind'; a challenging of taken-for-granted epistemological framings; an attention to what emerged from such encounters and a willingness to engage in what could be perceived as 'foolishness'. An important consideration of Bastian's project was *what emerges* if you ask the question of *how to hear the voice of the non-human other* rather than attempt to formulate generalised truth claims from what emerged.

D'Ignazio (2014) explored a research activity which involved walking along the course of a creek which runs through her neighbourhood, following its twists and turns, engaging specifically with the river and myriad participants which co-constitute it rather than approaching her neighbourhood from the perspective of roads and human habitation. She repeated the activity with a group of local school children, exploring with them the new perspective that walking/meandering with the creek afforded and the way this affected them.

Mukherjea and Bussolini (2014) explored listening to the voice of cats through a shared voluntary participation in yoga activities. The participation involved a physical closeness, a sharing of touch, a sense of the other's breathing, an attention to the present moment and an enhancement of the yoga activities[9] through these experiences. This attention to the present moment opened up the possibility of a friendship with non-human others based on Bergson's notion of *duration* rather than language. Such friendship in duration was built on a sharing of a '*vie en commun*' (life in common) in both space and time rather than for any utilitarian purpose. In their discussion of a 'life in common', Mukherjea and Bussolini (2014) also refer to Merleau-Ponty's (1962 [1954]) writing on the phenomenology of touch, internality and intertwining to explore the embodied enactment with the cats in the yoga session. Poulton (2014) provides a helpful discussion of 'phenomenological touch', highlighting how Merleau-Ponty offers a counterpoint to the reduction of others to 'mere objects of my internal perceptual consciousness' and how 'touch':

> opens up an *intertwining* [my italics] of inside and outside which is the overarching reality of the inter world in which we all exist and in which, through my body, I open myself up onto others in the same way they open themselves onto me.
>
> (Poulton 2014: 440)

The intertwining highlighted by Merleau-Ponty and Mukherjea and Bussolini (2014) is taken a step further by Braidotti and Haraway (as explored in Chapters 3 and 4). For Braidotti, in her posthuman framing, the encounter of the self with the non-human 'other' provides a space for exploring the multiple ways in which the other-than-human is already part of our human selves. For Haraway (2008: 5), we are 'beings-in-encounter where one is created in and through intra-and inter-action.

Puig de la Bellacasa's (2017) emphasis on 'touch', introduced in Chapter 3, can contribute to these discussions. Ticktin and Wijsman (2017: online) comment that for Puig de la Bellacasa touch (the haptic) is a way to shift away from abstraction and detachment and to engage ethically with what surrounds us in ways which can transform 'our thick present and the futures we want to co-create'. Touch understood in this framing is not a way of 'more accurately knowing a "real world"'. Rather it opens ethical possibilities for 'more involvement and commitment to it' (online).

The art installation *Ice Watch* (Eliasson 2014), a collaboration between the artist Olafur Eliasson and geologist Minik Rosing, provides opportunities for a first-being encounter with melting Arctic ice in ways which open possibilities for the ethical commitment proposed by Puig de la Bellacasa. Touch and attentive listening have an important role in the installation – enabling a direct, tangible experience of the reality of melting Arctic ice occurring in climate change. Drawing on behavioural psychology, Eliasson argues that such direct, embodied encounters have the potential to change human behaviour in ways that data and cognitive activity alone are unable to do (Jonze 2018).

The blocks, responsibly taken out of the waters of the Nuup Kangerlua Fjord (where they had already separated from the ice sheet), are presented in a clock formation in a prominent place open to the public.[10] Initially installed in Copenhagen to coincide with the publication of the United Nation's Intergovernmental Panel on Climate Change's (IPCC) *Fifth Assessment Report on*

Climate Change (2014), *Ice Watch* was then staged in Paris in 2015 to coincide with the United Nations COP21. In late 2018/early 2019, Ice Watch was set up in two locations in London to coincide with IPCC's 2018 publication, which warned that only 12 years remained to limit the worst effects of climate change. In all iterations of the installation (as pictures and films available on the internet document), touch, physical intertwining, listening and attentiveness through all the senses emerged as ways to encounter the ice. At press conference in London, Rosing commented:

> you'll notice they're all individuals, they are like beings. And they whisper to you, if you put your ear to them, you can hear the bubbles pop and you can see they're all different. And what they tell us is about a world that is different from the one that we have today. The air bubbles contain air that is fresh and clean and has half the CO_2 that we have in the atmosphere today. It brings us a message of tremendous change that is happening right now.
>
> (Luke 2018)

Intersubjective encounters with the wider natural world in which the authoritative position of *Anthropos* is deconstructed have implications for education. They extend the scope of the humanities, and of Western framings of education as a **human**ities subject, into new framings informed by both Western Posthumanist/posthuman[11] thinking and, equally importantly, by other ontological and epistemological positions such as the ones introduced and explored throughout this book. Such *revisioned* education can be a place 'to *play with the* "boundless, incalculable possibilities of life"' (Osberg 2015: 14) and subjectivities within it. It can be a place to challenge autonomous, rational, anthropocentric, androcentric and Eurocentric understandings of subject and create spaces for unique subjectivities to emerge – subjectivities which open possibilities for new futures. This does then raise the question of how such encounters can be encouraged in education and this is therefore the focus of the next section.

Implications and possibilities for education opened by intersubjective encounters within the wider natural world

As explored in the previous chapter in relation to first-person intersubjective encounters with other humans, a key aspect pf encouraging the possibility of engagement in first-being intersubjective encounter with the wider world is *the value* placed on such encounters. The challenge for teachers is to allow time for such encounters and also opportunities to reflect that human autonomy and domination over the wider natural world is not the only way to be. Such an approach is different from some kinds of 'learning in the outdoors', which emphasise what such learning can do for the child. For example, whilst the

UK based Natural Connections Demonstration Project (2012–2016) at Plymouth University emphasised a connected systems view of sustainability, its final report (Natural England 2016) focuses on outcomes which benefit the learner such as increased enjoyment and achievement; social skills across all curriculum areas and an improved sense of health and wellbeing. It pays little attention to the issue of questioning and repositioning of humans in relation to the wider natural world which has value and potential in and of itself.

In exploring how intersubjective first-being encounters can be encouraged in education, I draw on a range of ideas including 'slow education' and the value of pausing, Bennett on 'enchantment' and the Crex Collective on 'wild pedagogies'. These ideas can be used as starting points to encourage education as a 'time of regard for the world, of being present to it . . . of exposure effacing social subjectivities and orientations, a time filled with encounters' (Masschelein 2011: 1).

Slow education and the value of 'pausing'

'Slow education' is an educational response arising from the slow food movement whose naming is often attributed to the Italian food journalist Carlo Petrini. He posited the idea of 'slow food' as a response to his encounter with a newly opened MacDonald's restaurant at the Spanish Steps in Rome in 1986. The *slow food* movement stands in opposition to fast food produced through a 'mechanised production process easily replicated and scaled up across the world' (Harrison-Greaves 2016). Slow food emphasises local foods and production methods, responsiveness to local conditions and food prepared with care and enjoyed in a relaxed way with friends. The *slow school/education* movement (for example see Holt 2002) draws ideas from the slow food movement into educational settings. Slow schooling/education opens a breach in linear learning. Linear learning focuses on achieving as many tasks as possible and the acquisition of pre-set knowledge in step-by-step ordered linear time (expressed for example in school timetables). In contrast, slow education emphasises local knowledges and relationships; time for pausing; time for encountering; time for being-with and for learning *with* and *from* the other. It has potential to open moments for intersubjective first-being/first person encounters and the new subjectivities which, as I have argued in this and the previous chapter, can emerge in and through such intersubjective encounters.

The writer and poet Nan Shepherd conveys this sense of 'being with' in her work *The Living Mountain* (2011 [1977]). Through her writing, she demonstrates a somatic (bodily/whole body) attentiveness to the mountain in/with all its myriad participants and an attentiveness to the entanglement of our subjectivities with these myriad others. This is very different from approaching the mountain through a Western (Eurocentric) lens in which the mountain is 'brute and inert' – an object to be 'possessed', conquered, categorised, measured – resonating instead with other ways of being and knowing the world.

Bennett's thinking on *Enchantment* (2001) and *Vibrant Matter* (2010) can contribute to encouraging in educational settings the embodied experience and attentiveness Shepherd foregrounds, as I now explore.

Enchantment

Bennett proposes a re-awaking in Western (Eurocentric) thinking of how 'matter' constituting the world is, and always has been, 'vibrant' and 'enchanting'. For Bennett, 'to be enchanted is to be struck and shaken by the extraordinary that lives amid the familiar and everyday' (2001: 4). It is 'to participate in a momentary immobilising encounter' and 'be transfixed, spellbound' (5) in a 'moment of pure presence' (Fisher 1998: 131 cited in Bennett 2001: 5), which can open up 'a pleasurable feeling of being charmed' (5) and at the same time 'a more *unheimlich* (uncanny) feeling of being disrupted or torn out of one's default sensory-psychic-intellectual disposition' (5). Bennett comments that enchantment can 'hit us by surprise', but one can also foster certain strategies to cultivate it. Her suggestions, which can be drawn into educational settings, include 'giving greater expression to the sense of play', 'honing sensory receptivity to the marvellous specificity of things' and 'resisting the story of disenchantment of the modern world' (4). Enchantment is a process with potential to provoke 'new ideas, perspectives and identities' (6). To reinforce this point Bennett draws attention to the connection between the word 'enchantment' and the French verb *chanter* – to sing. The enchanted world 'sings', it generates refrains which 'turn back on themselves', 'open onto themselves' revealing:

> until-then unheard of potentialities, entering into other connections, setting [things] . . . adrift in the direction of other assemblages.
>
> (Deleuze and Guattari 1987: 349 cited
> in Bennett 2001: 6)

Bennett (2002) discusses Thoreau's encounters with 'wilderness' to explore 'enchantment' further. In many interpretations of Thoreau, 'the wild' is understood to be 'nature', especially nature untouched by humans. However, in Bennett's interpretation, Thoreau emphasises encounters in which 'wildness is that eccentric and recentering potential *within any object of experience*, once it is discerned in the right way . . . and thus ejected from the realm of everydayness' (Bennett 2002: xxii). In this line of thinking, enchanting encounters provide 'an impetus to engage with others as the individuals that they are' (xxiii) and 'energise one to affirm life and better appreciate living things' (xxiii).

Echoing Bergson's *élan vital* (the explosive force – due to an unstable balance of tendencies – which life bears within itself [1912: 103]), Bennett (2010: online) emphasises 'that enchantment is not so much a belief as it is an energetic current produced by the encounter between two sets of *active materialities*'. One can acknowledge and make connections here with Little Bear's

(2000: 77) discussion, explored in Chapter 2, of the Blackfoot tenet that 'Everything in existence, everything in creation, consists of energy waves' and bring the ideas into conversation, although it is important not to make simplistic comparisons.[12] Bennett highlights how one set of energy waves is 'congealed into a "self"' and one into what is often called the "objects"'. This 'energetic current' generated through encounters can be a source of ethical response. She comments:

> the *intensity* of the compound mood of enchantment (wonder/disturbance) could serve as one impetus to ethical action, insofar as it contributes the energy or motive force needed to render human bodies capable of jumping the gap between mere conviction that a course of action is good and the actual doing of the deed. What Spinoza called the 'joyful' affects are needed to energize a body called upon – by habit, sympathy, or reason – to love, forgive, treat with compassion or minimised harm to (an ontologically diverse range of) others. In short, I think that, under the right circumstances, the mood of enchantment, which entails the experience of the outside as *making a call*, can be an important part of ethics.
>
> (2010b: online)

Bennett's thinking on enchantment and ethics has some shared ideas (and also some differences) with the thinking of Bilgrami.[13] For Bilgrami, the source of living an 'unalienated life' is the liveliness of matter and the normative (ethical) demand that openness to and responding to this vibrancy places on each actor in the world. For both thinkers, encounters open possibilities of responding ethically to the other.

Education taking place outdoors has potential for enchantment, but it can also create barriers. Often it is very task-orientated, with an emphasis on the completion of activities and worksheets and engaging in processes which focus on classifying rather than attentive listening and openness to and 'learning with and from' the other. Such task-orientated learning was strikingly apparent to me on a walk I made between Malham Tarn and Malham Cove in Yorkshire, UK. This is an area well-known for its limestone pavements and the ecosystems these are part of. My walk coincided with a week popular in the UK school calendar for geography field trips. All along the route the children were busily engaged in tasks and worksheets – identifying, recording, collecting. There were moments for enchantment, for pausing to look around, for responding to and learning from other humans as well as the other-than-human world, but these were generally very limited. Like many visitors to Malham Cove, my walk culminated in a visit to the Janet's Foss 'fairy pond'. The fable tells that Jennett (Janet), Queen of the Fairies, lived in a cave at the falls. The setting of the falls and pond in a mossy glade adds to its 'fairy-tale' quality. The pond is approached along a mossy, wooded path which is indeed 'enchanting'. As I arrived at the pond I discovered it was busy with children of different ages in wellingtons, equipped with measuring sticks, tasked with wading in the water to record the depth in different areas. Gathering such data, it seems, is an aspect of the English National Geography Curriculum.

The geographer and teacher-educator Sharon Witt (2019) proposes a different approach to geography field trips. In contrast to Western thinking and methodologies focusing on weighing, measuring, collecting and analysing the world, Witt speaks of being a 'geographer without a map'. She suggests a collection of alternative verbs – 'lingering, mingling, exploring, noticing, connecting, sharing and greeting a multitude of things flourishing in relation'. These approaches can encourage the possibility of enchantment; processes of slow education and the being-with and somatic (whole body) attentiveness to the wider natural world as, for example, modelled by Nan Shepherd and Indigenous thinkers. They open potential for intersubjective first-being encounters with the wider natural world in and through which unique beings can emerge. The Crex Collective,[14] in their concept of *wild pedagogies*, also provide techniques which can encourage such emergent processes – as I now go on to explore.

Wild pedagogies

The Crex Collective (2018), who include I Hebrides as a co-author and active research participant, propose six touchstones for *wild pedagogies*: Nature as co-teacher; Complexity, the Unknown, and Spontaneity; Locating the Wild; Time and Practice; Socio-Cultural Change and Building Alliances and the Human Community. Wild pedagogies foreground and challenge how Eurocentric thinking imbues a consciousness of 'domestication, colonisation and destruction' (Jickling 2018: 40). The Collective use the term 'wild' in a 'doubled' sense. First, it is employed to indicate engagement with the 'wild' in ways which challenge Western (Eurocentric) destructive and unsustainable understandings of 'nature'. Second, it is used to indicate the necessity for *the 're-wilding' of the education process itself* towards a 'more demanding, rebellious, disruptive education for change' (Henderson 2018: 323), thereby contesting the increasing control over, and domestication of, educators and Westernised curricula. Whilst 'wild pedagogies' draw on the traditions of outdoor education, environmental education and deep ecology its distinctiveness lies in its foregrounding of the need to 'disrupt so much that is the *status quo* in society and how it is now served by education' (Jickling 2018: 35). It calls for a more 'humble humanity'; a re-examination of 'concepts of wilderness, wildness and freedom' and a forging of new relationships with 'places, landscapes, nature, more-than-human beings, and the wild' (35). This involves 'carefully listening to available voices', requiring 'actively de-centering the taken-for-granted human voice and re-centring of more-than-human voices' (35).

The six touchstones of wild pedagogy are not suggested dogmatically but rather as evolving 'points of departure' and places to return. In proposing 'nature as co-teacher' (Touchstone 1), the Collective (2018: 4) argue 'education is richer, for all involved, if the natural world and the many denizens that co-constitute places, are actively encountered, listened to, and taken seriously as part of the educative process'. Nature as co-teacher 'implies more than simply learning from the natural world; it includes learning with and through it as well; and thus, its myriad beings become active, fellow pedagogues' (2018: 5). The touchstone thus foregrounds the 'tremendous benefits to questioning the idea that a single human teacher should be at the centre of teaching and learning, and to expand[ing] consideration of what and who an educator is and might be' (5).

The second touchstone: *Complexity, the Unknown, and Spontaneity* argues, as I did in Chapter 2, that most often education 'is now conceptualised as the transference of a canonized body of knowledge from those who know to those who do not' (10) with 'what is considered knowledge and what is worth knowing' largely predetermined 'by those in control' (10). Yet, echoing Mouffe (2005, 2007), the Collective also note how 'even in this carefully constructed space, there lurk shadows and forgotten strands of complex interconnections' (10). To encourage *Complexity, the Unknown, and Spontaneity* the Collective propose taking risks, moving away from assumed ends and encouraging students to explore and be comfortable with the incompleteness of knowledge. Encounters have a role to play in this process: encounters which open possibilities for 'complex interconnection' with others, both human and other-than-human, opening spaces through which subjectivity emerges in ways 'which cannot be known' or 'predetermined and prescribed in advance' (The Crex Collective 2018: 10).

Touchstone Three, *Locating the Wild*, recognises the value of experiencing 'the wild' in remote areas and encouraging opportunities for students to have specific intersubjective encounters in such places. However, drawing on Næss, the Collective also emphasise that encounters with 'the wild' can be found everywhere: for example, a flower pushing through the corner of a playground. The teacher's role (and I would add that this initial impetus can also come from learners) is to notice, to respond to the plant with sympathy through both spoken and body language, encouraging a personal relationship, an 'enchantment' where one is 'struck and shaken by the extraordinary that lives amid the familiar and everyday' (Bennett 2001: 4). The fourth touchstone, *Time and Practice*, recognises that building such relationships with the wider natural world will, as in all relationships, take time, patience, discipline and practice. A starting point is for teachers to commit to this process themselves including making time to do this (an additional but worthwhile demand in the already overloaded reality of many teachers' lives) as well as encouraging students to do so.

Touchstone Five, *Socio-Cultural Change*, highlights that the way humans currently exist on the planet needs to change; that this requires change at a cultural level and that education has an important role to play in this process. For the Collective, '[e]ducation is always a political act, and we see wild pedagogues embracing the role of activists as thoughtfully as they can . . . telling a new *geostory* of a world in which all beings can flourish' (25–26). In reviewing this touchstone, Henderson (2018) recalls a wild pedagogy moment in his own practice. Students decided to place signs saying 'Entrance' over all the exit signs in their building, drawing attention to, and valuing, *entrance into* the wider natural world that lay beyond the walls: a reorientation, a wild act disrupting existing dominant Western conventions and mindsets and opening new ways to approach encounters there. Whilst students did not expect the signs to remain in place for long (for safety reasons), in that moment of resistance they felt free to celebrate the other-than-human and to reposition the human in relation to these others. Henderson comments that even 20 years later, students recalled it as significant: an act which shaped *who* they became as unique actors in the world.

In their final touchstone, *Building Alliances and the Human Community*, the Collective emphasise that 'the colonial ethos of resource extraction is not separate from but is yet another shade of the many hierarchies of dominance that exist amongst humans': a dominance which also contributed/contributes to the 'unhumaning' of black and Indigenous people which occurred/occurs as part of the European project of colonisation (Jackson 2013). Wild pedagogues are therefore committed to 'seeking alliances and building community with others not only in the environmental world but across all people and groups concerned with justice' (The Crex Collective 2018: 31). Wild pedagogues recognise their interdependency with communities and the importance of the presence and role of such communities in wild pedagogies. To encourage the building of such alliances the Collective propose encouraging oneself and one's students to notice and respond to colonising perspectives both towards other humans and the wider natural world, to 'lean in, and listen' and to explore and encourage acts of resistance by exploited/colonised humans and other-than-humans. Andreotti's (2012, 2015, 2016) HEADS UP model introduced in Chapter 5 can assist in this process.

As these various discussions highlight, opportunities to engage in intersubjective encounters exist all around us if there are moments to pause, to take the time to be attentive, to approach the other with sympathy. However, educators may also wish to arrange specific opportunities for encounters with the wider natural world in conjunction with organisations such as farms, countryside and wildlife centres and forest schools, as the following case studies explore.

Case studies: supporting encounters within the wider natural world in educational settings

Encounters at Jamie's Farm, UK

Encounters within the wider natural world are a key aspect of Jamie's Farm (2020a) (introduced in the previous chapter), whether this be the impact of the open vista across the fields and views of a wide horizon, unavailable to children living in the inner-city, or encounters with specific animals, plants and places. Some encounters arise through animal care, ranging from feeding and cleaning-out to helping with lambing. The animals have no negative preconception of the young people involved in their care and also rely on the participants for their wellbeing.[15] The young people involved in the project are unused to being approached in this way. The trust placed in them opens space to engage, space for spontaneity and experiencing the unknown and the possibilities of emergence (recalling wild pedagogies' touchstone two), of new ways to be and be together – ways which continue when the participants then return to their schools.

Horse therapy is also an important aspect of the work of Jamie's Farm.[16] Horses are very affected by the mood of those around them, and, responding to this, students can develop the relaxed state needed to help the horse relax. A horse is a powerful animal, and at first students can be apprehensive. However, engaging in an encounter with a particular horse on a particular occasion (the horse is a co-teacher alongside the therapist) can help to develop trusting, affectionate relationships. This has the potential to open the emergence of new, often surprising, ways for a student to be in the wider world: changes which then also surprise their teachers when they return to their school. Jamie's Farms are residential but there are opportunities for day visits to other projects such as countryside and urban farms. Some centres bring encounters into schools and colleges. The Caenhill Countryside Centre, Wiltshire UK provides both such opportunities.

Encounters at Caenhill Countryside Centre, Wiltshire UK

Caenhill Countryside Centre is a 'community not-for-profit organisation working with communities to bring agriculture and horticulture to children and young people by providing courses and hands-on experiences set on a farm' (Caenhill Countryside Centre: online). It also provides opportunities for the development of rural and conservation/heritage skills and crafts and woodwork and metal work. In addition, they take activities into educational establishments. However, the centre's work goes further than learning skills, as its website tagline '*Caenhill: Touching Hearts and Lifting Souls*' indicates. As a working farm they produce crops for market and to feed their animals. However, they never send their animals to market, emphasising on their website (echoing wild pedagogies' touchstone one) that all animals 'are part of our education team and, like us, work with young people's education and wellbeing'. Through participating in the work and activities of the Centre, and encountering animals there as co-teachers in first-being intersubjective encounters, students develop practical skills and awareness of how their actions can lead to safe and caring outcomes for others and their own valuable place in the world. The Centre encourages a trusting, harmonious and fun atmosphere. No animals are chased (by people or other animals), and they do not feel the need to move away from humans. All of these factors contribute to an atmosphere in which first-being intersubjective encounters with animals and other participants in the wilder natural world open possibilities for students to be and become in new ways in the world.

There are other opportunities to spend time and 'be' at farm, countryside and conservation centres. In countries in the global North, visiting working farms is made possible through organisations such as The Country Trust[17] in the UK and the Landbrug & Fødevarer (Agriculture & Food Council) in Denmark. There are also opportunities provided by organisations connected to, or informed by, the international Forest School Movement. Sackville-Ford (cited in Lightfoot 2019: online) identifies that forest schools adopt an approach which is 'slightly subversive' – a 'rewilding' of the education process itself. It sets children free to be 'guided by their own curiosity rather than completing tasks set by the teacher' and to learn through 'play and discovery, collaboration and risk taking, climbing trees and using knives'. However, it is important to note that not all education labelling itself as 'forest schooling' is in line with these ideas (for example, see discussion in Sackville-Ford and Davenport 2019).

If the educational approaches explored here are to open spaces for intersubjective first-being encounters, the possibility of natality and emergence of new ways of knowing and being together, some things are key. There need to be opportunities for 'slow education', for pausing and attentiveness, for tenderness, for enchantment, for 'lingering, mingling, exploring, noticing, connecting, sharing and greeting a multitude of things flourishing in relation' (Witt 2019). Specific learning tasks may exist, and these can be an important part of multifaceted educational processes. However, these should not exclude moments for natality, emergence and a valuing of intersubjective first-being encounters when they do occur. Whilst the emphasis here has been on valuing intersubjective first-being encounters, it is also important to be aware of the potential danger of romanticising encounters and also appearing to suggest they are the only way to approach our relationship with the wider natural world – issues I now explore.

Exploring ways to avoid romanticising encounters and finding ways to respond to others we are unable to encounter

In order not to 'romanticise' the wider natural world, it is necessary to recognise and explore the possibility of harsher and also more 'neutral' encounters. Plumwood (1995) relates her experience of being attacked by a crocodile, and the realisation that for the crocodile she was food. In the encounter she came to understand that this was not a personal attack, but no less devastating for that. This harsh first-being encounter led her to re-evaluate her own relationship with the wider natural world and to consider that the way we treat animals who are food to us could be understood as just as harsh. It led her to engage with the possibility of encountering the wider natural world in a different way.

Whilst it may not be possible to include harsh first-being encounters in education, it is possible to reflect on ones that students may have experienced, discuss their impact and explore how these have become part of who they are and how they act. Although a very different encounter from Plumwood's, as a young child I experienced a challenging first-being intersubjective encounter of my own, which I can recall vividly over 50 years later. I include it here as a 'word picture'. I expect that readers of this book have their own vivid recollections of first-being encounters which remain with them, forming part of who they are as actors in the world, and can also encourage students to reflect in this way.

> On a walk with our mother my sister and I saw a distressed, disfigured rabbit writhing in extreme pain and with red swollen eyes. My mother called us away, explaining that the rabbit had Myxomatosis and was dying.[18] She explained that this was a disease deliberately introduced to reduce the rabbit population because of concerns about rabbits eating crops. I can now reflect on the complex issues involved. Yet I can also still vividly recall the encounter itself and also my inability, at first, to comprehend my mother's words. I felt I must be misunderstanding my Mother's explanation as it did not seem possible to me that such a disease and such visible suffering could have been deliberately encouraged. I was shocked when my mother explained it to me again and I realised my initial understanding was correct. This made me reconsider my trust in adults and question the ways adults sometimes choose to treat the wider natural world. The encounter with this rabbit, on this occasion evoked an ethical response in me which started a new way for me to frame the world. One could argue that I am now layering onto this experience ideas of framing and ethics. However, whilst I certainly did not have the vocabulary at the age of six to express my thinking in the way explored here I would argue that I was capable of responding in an ethical way nonetheless – a way that is still part of my own material historicity.

This encounter with the dying rabbit is not included as a criticism of those in agriculture who are engaged in their own first-hand encounters with the wider natural world, and who face challenges and difficulties of their own. Rebanks (2015), a farmer and writer in the Lake District UK, describes how he came to realise that farmers' love of the natural environment and their livestock is very different from the experience of the Romantic and modern-day poets and artists. It draws on a very different kind of first-hand encounter but is valuable and felt with passion. The writer J.S. Collis (1975: 11) touches on this difference as he explains his request to work in agriculture rather than be allocated a desk job in the Second World War:

> I had hitherto regarded the world from the outside, and I wished to become more involved in it. . . . I gained the opportunity to become thoroughly implicated in the fields instead of being merely a spectator of them.

In his writing, Collis highlights the often-harsh realities and very hard work of the agricultural world but also the sense of ecological interconnectedness and contentment he experienced in encounters there.

Bastian (2019) highlights another issue in relation to foregrounding encounters with the wider natural world, namely there are others one can never encounter because, for example, they are beyond where a particular individual can travel or even beyond anyone's ability to travel. This line of thinking led her to explore 'whalefall' and all that depends upon it: a hidden ecosystem which no-one encounters. These ecosystems develop when whales die and fall to the ocean floor, with a single large whale supporting myriad non-human others for an estimated period of up to 75 years. Thus, the decline in whales significantly threatens these hidden others who are now being brought to human attention through the use of unmanned submarines. In highlighting that intersubjective encounters with these threatened others is not possible, Bastian draws attention to the need that 'encounters' cannot be the sole approach to engaging with others, particularly in the context of the issue of human exploitation of the wider natural world. Here the ideas of Bennett (2010b: online) can be drawn in. As cited earlier in the chapter, Bennett proposes that 'the *intensity* of the compound mood of enchantment (wonder/disturbance)' which is generated in first-being encounters can contribute to a wider ethical response and action, as enchantment can generate 'the energy or motive force needed to render human bodies capable of jumping the gap between mere conviction that a course of action is good and the actual doing of the deed'. Thus, intersubjective encounters which students *do* experience can encourage the 'intensity' which can give rise to ethical responses to others whom students cannot experience directly.

The ideas and examples explored in this and the previous chapters point towards the ways in which first-being/first-person intersubjective encounters have potential to be ethical encounters, raising the questions of how Arendt's concept of forgiveness and mutual promising could contribute. This is therefore the issue I now examine.

Intersubjective encounters within the wider natural world: a role for forgiveness and mutual promising

Arendt developed her conception of forgiveness and mutual promising as a way to respond ethically to issues arising in specific encounters between humans. It is challenging to see whether and how these ideas can fruit-fully be explored in relation to encounters between humans and the wider natural world. Questions arise around whether an animal, plant or a rock can forgive and make mutual promises, and how these actions could be understood in ways not limited to dominant human definitions. Asserting that such ethical responses are possible pushes at the boundaries of the way many (but not all) people, especially those in the Global North, under-stand the world. For example, some question whether the agency and reci-procity needed for forgiveness and mutual promising are possible in such encounters. However, the various approaches to encountering, hearing and entering into reciprocity with non-human others discussed in this chapter suggest that this is a *possibility*. Whilst Arendt's conception cannot be trans-posed in a *simplistic* way to encounters within the wider natural world, it can have meaning there.

Arendt emphasises how moving from encounter to encounter can encourage the emergence of new forms of '"general"[19] understanding' including ethi-cal understanding, and this includes new ways to think about the reciprocity involved in forgiving and mutual promise-making. For example, whilst a tree does not make a promise in the way humans do, encountering and being open to the idea of entering into a reciprocal relationship with a specific tree in an 'I-thou' relationship (as proposed, for example by Buber) opens opportuni-ties to rethink what forgiveness and mutual promising can mean in such an encounter. For example, through embodied faithfulness and despite harm that 'I' have contributed, e.g. through pollution, *this* tree commits to sustaining life, including my own.

In this chapter and previous chapters I have explored how first-person/first-being encounters open potential for shifts from static or preconceived con-cepts of human subjectivity. Through radically open, democratic and dynamic processes the self can emerge intersubjectively, 'never outgrowing its open malleable ego boundaries' (Keller 1986: 134). Such processes open potential for something new, something which did not exist before and could not be predicted. Education which encourages and values first-person/first-being encounters, whilst also acknowledging and finding ways to respond to the ethical risks inherent in such processes, opens possibilities for emergence of new ways to be and become in the world and to act and respond to the social and environmental challenges of this and future eras.

Notes

1 As discussed in Chapter 7, the case studies here are projects designed to support students who are 'not doing well' academically and/or emotionally in mainstream UK settings still strongly influenced by Western (Eurocentric) framing of the autonomous rational individual. They provide ideas which can also benefit all students. Whilst examples can help to elucidate theory they can also inform future theorising in a 'patterning exchange' between theory and practice.

2 Lyvers (1999) focuses on the issue of animal subjectivity, but in this chapter I extend ideas to incorporate all parts/participants of the wider natural world.

3 As introduced in Chapter 6, ecocide is such extensive damage to, destruction of or loss of ecosystem(s) that the very survival of that ecosystem and all who form part of it is threatened.

4 To respond to this inability to listen Blenkinsop *et al.* (2017) argue for extending Memmi's (1991[1957]) concept of 'listening first' to listening to the wider natural world.

5 The ability of genes to change position on chromosomes, a process in which a transposable element is removed from one site and inserted into a second site in the DNA (Medicinenet.com 2016).

6 Holdrege (2005) provides examples of how a tree is responsive to its environment, with each seed developing into a unique tree. He argues that unquestioning reliance on theories of competition and survival of the fittest to understand the wider natural world is insufficient and invites an opening up of other ways of understanding.

7 For example, see Little Bear (2000, 2009, 2011, 2016), Bawaka Country *et al.* (2016), KARI-OCA 2 Declaration (2012), De Line (2016) and discussion in Chapters 2 and 3.

8 The workshop was a collaboration between Plymouth University UK, the Authority Research Network and the Public Science Project at City University of New York.

9 The researchers held yoga sessions in a space shared with the cats at their research centre and discovered that cats were voluntary, active and frequent participants in these sessions. They then undertook historical research and found that yoga positions in some cultures drew on feline movements.

10 Eliasson, aware of the environmental impact of transporting ice and for each installation, asked the NGO Julie's Bicycle to calculate the project's carbon footprint. These calculations showed that Eliasson's Ice Watch Paris produced 30 tonnes of carbon dioxide which is the CO_2e equivalent of 30 people flying return from Paris, France, to Nuuk, Greenland. Eliasson believes that this carbon footprint is acceptable, given the many thousands of people who have an opportunity through the installations to have a direct encounter with Arctic ice, opening possibilities for new ways of acting necessary to combat climate change (Phaidon 2019).

11 Whilst the focus in this chapter is on intersubjective encounters in education between the self and other participants of the wider natural world which appear 'exterior' to the self, I recognise that in posthuman thinking the boundaries between human bodies and the other-than-human is blurred since human bodies are themselves already constituted in myriad ways with the other-than-human.

12 See *Rosiek et al.* (2020) for further discussion.

13 Bennett acknowledges Bilgrami's thinking on enchantment but identifies certain key differences such as the way that Bilgrami emphasises the re-enchantment of a secular world, whilst Bennett emphasises that the world has always been and remained vibrant and enchanting. The *source* of the ethical call that enchanting encounters generate also differs. See Bennett (2010b) for further discussion of these points.

14 The Crex Collective members are I Hebrides, Ramsey Affifi, Sean Blenkinsop, Hans Gelter, Douglas Gilbert, Joyce Gilbert, Ruth Irwin, Aage Jensen, Bob Jickling, Polly Knowlton Cockett, Marcus Morse, Michael De Danann Sitka-Sage, Stephen Sterling, Nora Timmerman and Andrea Welz. I Hebrides is included as a co-author and active research participant. 'Crex crex is the taxonomical name given the Corncrake. We have chosen this bird because it was an important collaborator in this project and because its onomatopoeic name beautifully mirrors its call – a raspy crex crex' (Crex Collective 2018: 1).

15 Jamie's Farm acknowledge the issue of animal welfare and raising animals for meat. They do not shy away from these issues but instead aim to help students understand food production, the ethical issues this involves and the possibility of making informed choices about the food one eats.

16 Horse therapy can be explored in more detail in a range of literature, for example see Scharff 2017.
17 The Country Trust's aim is to reconnect children with the working countryside (The Country Trust, online).
18 Myxomatosis is a poxvirus which originated in the American continent (Arthur and Louzis 1988). It was brought into France deliberately by a landowner who wanted to control the rabbit population on his land (Bartrip 2008). The disease spread quickly through mainland Europe and made its appearance in the UK in 1953. It was not official government policy to encourage the spread of the disease, and in the end, legislation was brought in to outlaw this practice, partly due to public pressure.
19 See discussion in Chapter 5 of Arendt's thinking on how general understanding can emerge through specific encounters rather than starting with a general understanding under which particular experiences are then subsumed.

References

Andreotti, V. (2012) Editor's preface: HEADS UP. *Critical Literacy: Theories and Practices*, 6(1): 1–3. Available at: www.oregoncampuscompact.org/uploads/1/3/0/4/13042698/andreotti_-_preface_-critical_literacy_org_-_headsup__1_.pdf [Accessed 23.1.2020].

Andreotti, V. (2015) Global citizenship education otherwise: Pedagogical and theoretical insights. In A. Abdi, L. Shultz and T. Pillay (eds.) *Decolonising global citizenship education*. Rotterdam: Sense Publishers, pp. 221–230.

Andreotti, V. (2016) The educational challenge of imagining the world differently. *Canadian Journal of Development Studies/Revue canadienne d'études du développement*, 37(1): 101–112.

Arendt, H. (1974 [1958]) *The human condition*. Chicago IL: University of Chicago Press.

Arendt, H. (2006 [1961]) What is authority? In *Between past and future: Eight exercises in political thought*. London: Penguin.

Arthur, C.P. and Louzis, C. (1988) A review of myxomatosis among rabbits in France. *Revue scientifique et technique (International Office of Epizootics)*, 7(4): 959–976.

Bartrip, P.W.J. (2008) Myxomatosis in 1950s Britain. *Twentieth Century British History*, 19(1): 83–105. doi: 10.1093/tcbh/hwm016

Bastian, M. (2014) Multi species methods: Participatory research and the more- than-human. *Democracy's Non-Human and Non-Living "Others" International Workshop, CUNY Public Science*, 2nd May. Available at: www.authorityresearch.net/participations-others-audio-recordings.html [Accessed 8.11.2016].

Bastian, M. (2019) Whale falls, suspended ground, and extinctions never known. *Environmental Humanities Seminar Series, Public Lecture*, 11th December.

Bawaka Country, Wright, S., Suchet-Pearson, S., Lloyd, K., Burarrwanga, L., Ganambarr, R., Ganambarr-Stubbs, M., Ganambarr, B., Maymuru, D. and Sweeney, J. (2016) Co-becoming Bawaka: Towards a relational understanding of place/space. *Progress in Human Geography*, 40(4): 455–475.

Beavan, C. (2016) *The genius of Turner: The making of the industrial revolution*. Fell, R. and Beavan, C. (producers), 12th September. London: British Broadcasting Company (BBC2).

Bennett, J. (2001) *The enchantment of modern life: Attachments, crossings and ethics*. Princeton, NJ: Princeton University Press.

Bennett, J. (2002) *Thoreau's nature: Ethics, politics and the wild*. Lanham, MD: Rowman & Littlefield.

Bennett, J. (2010) On the call from the outside. *The Immanent Frame*. Available at: https://tif.ssrc.org/2010/08/18/on-the-call-from-outside/ [Accessed 4.1.2020].

Bergson, H. (1912) *Creative evolution*. Mitchell, A. (tr.). London: Palgrave Macmillan.

Blenkinsop, S., Affifi, R., Piersol, L. and De Danann Sitka-Sage, M. (2017) Shut up and listen: Implications and possibilities of Albert Memmi's characteristics of colonisation upon the 'natural world'. *Studies in Philosophy and Education*, 36(3): 349–365.

Bonnett, M. (2000) Environmental concerns and the metaphysics of education. *Journal of the Philosophy of Education*, 34(4): 591–602.

Bonnett, M. (2002) Education for sustainability as a frame of mind. *Environmental Education Research*, 8(1): 9–20. doi: 10.1080/13504620120109619

Bonnett, M. (2004) *Retrieving nature: Education for a post-humanist age*. London: Wiley.

Borgmann, A. (1992) *Crossing the postmodern divide*. Chicago, IL: University of Chicago Press.

Braidotti, R. (2013b) *The posthuman*. Boston, MA and Cambridge: Polity Press.

British Broadcasting Company (BBC) Media Centre (2019) *BBC One announces brand new series on plants, the green planet*. Available at: www.bbc.co.uk/mediacentre/latestnews/2019/the-green-planet [Accessed 18.12.2019].

Buber, M. (1958 [1923]) *I and thou*. Kaufmann, W. (tr.). New York: Scribner.

Caenhill Countryside Centre (online) *Caenhill countryside centre home page*. Available at: www.caenhillcc.org.uk/ [Accessed 7.4.2020].

Collis, J.S. (1975) *The worm forgives the plough*. Harmondsworth: Penguin.

The Country Trust (online) *Because the countryside can change lives*. Available at: www.countrytrust.org.uk/ [Accessed 8.1.2020].

Crex Collective (2018) Six touchstones for wild pedagogies in practice. Pre-publication copy of chapter 5. In R. Jickling, S. Blenkinsop, N. Timmerman and M. Sitka-Sage (eds.) *Wild pedagogies: Touchstones for re-negotiating education and the environment in the Anthropocene* (Palgrave studies in educational futures). New York: Palgrave Macmillan. Available at: www.researchgate.net/publication/328615912_Chapter_5_Six_Touchstones_f or_a_Wild_Pedagogy [Accessed 7.1.2020].

Deleuze, G. and Guattari F. (1987) *A thousand plateaus: Capitalism and schizophrenia*. Minnesota, MN: University of Minneapolis Press.

De Line, S. (2016) All my/our relations: Can posthumanism be decolonised? *Open! Platform for Art, Culture & the Public Domain*. Available at: www.onlineopen.org/all-my-our-relations [Accessed 23.1.2018].

Deloria, V. (1999) *Spirit and reason: The Vine Deloria Jnr. Reader*. Golden, CO: Fulcrum.

Derrida, J. (2002) The Animal that therefore I am (more to follow). Wills, D. (tr.). *Inquiry*, 28(2): 369–418.

Derrida, J. (2008) *The Animal that therefore I am (more to follow)*. New York: Fordham University Press.

D'Ignazio, C. (2014) The babbling brook. Democracy's non-human and non-living "others" international workshop. *CUNY Public Science*, 2nd May. www.authorityresearch.net/participations-others-audio-recordings.html [Accessed 8.11.2016].

Eliasson, O. (2014) *Ice watch*. Available at: https://olafureliasson.net/archive/artwork/WEK109190/ice-watch [Accessed 19.3.2020].

Fisher, P. (1998) *Wonder, the rainbow and the aesthetics of rare experiences*. Cambridge, MA. Harvard University Press.

Ground, I. (2013) *Listen to the lion: Wittgenstein and animal minds*. Tenth British Wittgenstein Society Lecture, 14th May. Available at: www.academia.edu/9298683/Listen_to_the_Lion__Wittgenstein_a nd_Animal_Minds [Accessed 19.3.2020].

Haraway, D. (2008) *When species meet*. Minneapolis and London: University of Minnesota Press.

Haraway, D. (2016) *Staying with the trouble: Making kin in the Chthulucene*. Durham, NC: Duke University Press.

Harrison-Greaves, J. (2016) Slow education leads to rich and balanced learning. *BERA Blog*. Available at: www.bera.ac.uk/blog/slow-education-leads-to-rich-and-balanced-learning [Accessed 6.1.2020].

Henderson, R. (2018) Review of 'Wild pedagogies: Touchstones for re-negotiating education and the environment in the Anthropocene' by B. Jickling, S. Blenkinsop, N. Timmerman, M. De Dannan Sitka-Sage (Editors.). *Journal of Outdoor and Environmental Education*, 21: 331–335. doi: 10.1007/s42322-018-0025-6

Holdrege, C. (2005) The forming tree. *In Context*, 14(Fall). Available at: http://natureinstitute.org/pub/ic/ic14/trees.htm [Accessed 10.1.2017].

Holt, M. (2002) It's time to start a slow school movement. *Phi Delta Kappan*, 8(4): 264–271.

Honig, B. (1995) Toward an agonistic feminism: Hannah Arendt and the politics of identity. In B. Honig (ed.) *Feminist interpretations of Hannah Arendt*. Philadelphia: Pennsylvania State University Press, pp. 135–166.

Intergovernmental Panel on Climate Change (IPPC) (2014) *Fifth assessment report*. Available at: www.ipcc.ch/ [Accessed 16.6.2014].

Jackson, Z.I. (2013) Review: Animal: New directions in the theorisation of race and post-humanism, reviewed work(s): HumAnimal: Race, law, language by Kalpana Rahita Seshadri; The birth of a jungle: Animality in progressive-era U.S. literature and culture by Michael Lundblad; Animacies: Biopolitics, racial mattering, and queer affect by Mel Y. Chen. *Feminist Studies*, 39(3): 669–685.

Jamie's Farm (2020a) *About us: Our journey*. Available at: https://jamiesfarm.org.uk/about-us/our-journey/ [Accessed 18.3.2020].

Jickling, B. (2018) *On wilderness*. In R. Jickling, S. Blenkinsop, N. Timmerman and M. De Danann Sitka-Sage (eds.) *Wild pedagogies: Touchstones for re-negotiating education and the environment in the Anthropocene* (Palgrave studies in educational futures). New York: Palgrave Macmillan, pp. 23–50.

Jonze, T. (2018) Iceberg's ahead! Olafur Eliasson brings the frozen fjord to Britain. *The Guardian*, 11th December. Available at: https://www.theguardian.com/artanddesign/2018/dec/11/icebergs-ahead-olafur-eliasson-brings-the-frozen-fjord-to-britain-ice-watch-london-climate-change

KARI-OCA 2 Declaration (2012) *KARI OCA 2 declaration: Indigenous peoples global conference on RIO + 20 and mother earth*. Accepted by Acclamation, Kari-Oka Village, at Sacred Kari-Oka Púku, Rio de Janeiro, Brazil, 17th June. Available at: https://wrm.org.uy/other-relevant-information/kari-oca-2-declaration-indigenous-peoples-global-conference-on-rio-20-and-mother-earth/ [Accessed 17.12.2019].

Keller, C. (1986) *From a broken web: Separation, sexism and self*. Boston, MA: Beacon Press.

Klein, N. (2015) *This changes everything: Capitalism vs. the climate*. London: Penguin.

Klein, N. [narrator and original book writer] and Lewis, A. [film writer and director] (2015) *This changes everything*. Klein Lewis and Louverture Film Production: New York.

Kuletz, V. (1998) *The tainted desert: Environmental and social ruin in the American West*. New York: Routledge.

Lightfoot, L. (2019) Forest schools: Is yours more a marketing gimmick than an outdoors education? *The Guardian*, 25th June. Available at: www.theguardian.com/education/2019/jun/25/forest-schools-more-marketing-than-outdoor-education [Accessed 8.1.2020].

Little Bear, L. (2000) Jagged world views colliding. In M. Batisse (ed.) *Reclaiming Indigenous voice and vision*. Vancouver, BC: University of British Columbia Press. Also available at:

www.learnalberta.ca/content/aswt/worldviews/documents/jagged_worldviews_ collid-ing.pdf [Accessed 14.10.2019].

Little Bear, L. (2009) *Naturalising Indigenous knowledge synthesis paper*. Aboriginal Learn-ing Centre, University of Saskatchewan. Available at: www.afn.ca/uploads/files/education/21._2009_july_cclalkc_leroy_littlebear_n aturalizing_indigenous_knowledge-report.pdf [Accessed 13.10.2018].

Little Bear, L. (2011) *Native science and Western science: Possibilities for collaboration*. Lec-ture on 4th of March at Arizona State University. Available at: www.youtube.com/watch?v=ycQtQZ9y3lc [Accessed 15.11.2018].

Little Bear, L. (2016) *Blackfoot metaphysics 'waiting in the wings'*. Congress of the Humanities and Social Sciences Big Thinking Lecture, 1st June. Available at: www.youtube.com/watch?v=o_txPA8CiA4 [Accessed 18.11.2018].

Loidolt, S. (2016) Hannah Arendt's conception of actualized plurality. In D. Moran and T. Szanto (eds.) *Phenomenology of sociality: Discovering the 'we'*. London: Routledge pp. 42–55.

Luke, B. (2018) Olafur Eliasson's latest work is melting away on the bank of the Thames in London. *The Art Newspaper*, 11th December. Available at: www.theartnewspaper.com/news/ice-watch-olafur-eliasson [Accessed 18.12.2019].

Lyvers, M. (1999) Who has subjectivity? *Psyche: An Interdisciplinary Journal of Research on Consciousness*. Available at: http://epublications.bond.edu.au/hss_pubs/12 [Accessed 3.2.2016].

Masschelein, J. (2011) *Experimentum scholae*: The world once more . . . but not (yet) finished. *Studies in the Philosophy of Education*, 30(5): 529–535. doi: 10.1007/s11217-011-9257-4

Massumi, B. (2014) *What animals teach us about politics*. Durham, NC: Duke University Press.

Medicinenet.com (2016) *Definition of transposition, genetics*. Available at: www.medicinenet.com/script/main/art.asp?articlekey=19485 [Accessed 13.1.2017].

Memmi, A. (1991 [1957]) *The coloniser and the colonised*. Introduction by Jean-Paul Sartre, afterword by Susan Gilson. Greenfeld, H. (tr.). Boston, MA: Beacon Press.

Merleau-Ponty, M. (1962 [1954]) *The phenomenology of perception*. Smith, C. (tr.). London: Routledge and Kegan Paul.

Mouffe, C. (2005) *The democratic paradox*. New York: Verso.

Mouffe, C. (2007) Artistic activism and agonistic spaces. *Art and Research: A Journal of Ideas, Contexts and Methods*, 1(2). Available at: www.artandresearch.org.uk/v1n2/mouffe.html [Accessed 9.2.2015].

Mukherjea, A. and Bussolini, J. (2014). La Vie en commun: Cat yoga. *Democracy's Non-human and Non-living "Others" International Workshop, CUNY Public Science*, 2nd May. Available at: www.authorityresearch.net/participations-others-audio-recordings.html [Accessed 8.11.2016].

Natural England (2016) *Natural connections demonstration project, 2012–2016: Final report* (Natural England commissioned report NECR215). Available at: http://publications.naturalengland.org.uk/publication/6636651036540928 [Accessed 22.11.2016].

Noddings, N. (2013) *Caring: A relational approach to ethics and moral education*. Oakland: Uni-versity of California Press.

Osberg, D. (2015) Learning, complexity and emergent (irreversible) change. In E. Har-greaves (ed.) *Sage handbook of learning*. London: Sage, pp. 23–40.

Pedersen, H. (2010) Is 'the posthuman' educable? On the convergence of educational phi-losophy, animal studies, and posthumanist theory. *Discourse: Studies in the Cultural Politics of Education*, 31(2): 237–250. doi: 10.1080/01596301003679750

Phaidon (2019) *Olafur Eiasson's ice watch will give you a climate change chill*. Available at: https://uk.phaidon.com/agenda/art/articles/2018/december/11/olafur-eliassons-ice-watch-will-give-you-a-climate-change-chill/ [Accessed 19.12.2019].

Plumwood, V. (1995) Human vulnerability and the experience of being prey. *Quadrant*, 29(3): 29–34.

Plumwood, V. (2001) *Environmental culture: The ecological crisis of reason*. London: Routledge.

Poulton, J.L. (2014) Fairbairn and the philosophy of intersubjectivity. In G.S. Clarke and D.E. Scharff (eds.) *Fairbairn and the object relations tradition*. London: Karnac Books, pp. 431–444.

Prempeh, (2008) The anticapitalism movement and African resistance to neoliberal globalization, neoliberalism and globalization in Africa. In J. Mensah (ed.) *Neoliberalism and globalisation in Africa: Contestations from the embattled continent*. New York: Palgrave Macmillan, pp. 55–69.

Puig de la Bellacasa, M. (2017) *Matters of care: Speculative ethics in more than human worlds*, 3rd Edition. Minneapolis: University of Minnesota Press.

Rancière, J. (2010) Who is the subject of the rights of man? In J. Rancière (ed.) *Dissensus: On politics and aesthetics*. London: Continuum, pp. 70–83.

Rebanks, J. (2015) *The shepherd's life: A tale of the Lake District*. London: Allen Lane.

Rosiek, J., Snyder, J. and Pratt, S. (2020) The new materialisms and Indigenous theories of non-human agency: Making the case for respectful anti-colonial engagement. *Qualitative Inquiry*, 26(3–4): 331–346.

Sackville-Ford, M. and Davenport, H. (eds.) (2019) *Critical issues in forest schools*. London: Sage.

Samuel, S. (2019) Forget GDP – New Zealand is prioritizing gross national well-being. *Vox*, June 8th. Available at: www.vox.com/future-perfect/2019/6/8/18656710/new-zealand-wellbeing-budget-bhutan-happiness [Accessed 13.12.2019].

Scharff, C. (2017) The therapeutic value of horses. *Psychology Today*, 23rd August. Available at: www.psychologytoday.com/gb/blog/ending-addiction-good/201708/the-therapeutic-value-horses [Accessed 15.1.2020].

Shepherd, N. (2011 [1977]) *The living mountain*. Edinburgh: Canongate Books.

Singh Bindra, P. (2019) *The vanishing: India's wildlife crisis*. New Delhi: Penguin, Random House India.

Spivak, G. (1993) *The post-colonial critic: Interviews, strategies, dialogues*. New York: Routledge.

Ticktin, M. and Wijsman, K. (2017) *Review: Maria Puig de la Bellacasa – matters of care: Speculative ethics in more than human worlds*. Available at: www.hypatiareviews.org/reviews/content/337 [Accessed 21.1.2020].

Todd, Z. (2016) An Indigenous feminist's take on the ontological turn: 'Ontology' is just another word for colonialism. *Journal of Historical Sociology*, 29(1): 4–22.

United Nations Intergovernmental Science-Policy Platform (IPBES) (2019) *Biodiversity and ecosystem services assessment*. Available at: https://www.un.org/sustainabledevelopment/blog/2019/05/nature-decline-unprecedented-report/ [Accessed 16.12.2019].

Walsh, B.J., Karsh, M.B. and Ansell, N. (1994) Trees, forestry and the responsiveness of creation. *Cross Currents*, 44(2): 149–163. Available at: www.crosscurrents.org/trees.htm [Accessed 16.12.2019].

Watts, V. (2013) Indigenous place-thought and agency amongst humans and non humans (first woman and sky woman go on a European world tour!). *Decolonization: Indigeneity, Education and Society*, 2: 20–34.

Weil, K. (2008) Review of Derrida, Jacques, 'The animal that therefore I am'. *H-Animal, H-Net Reviews*, October. Available at: http://www.h-net.org/reviews/showrev.php?id= 22808 [Accessed 7.4.2016].

Witt, S. (2019) *Becoming lost in relational, democratic geographical fieldwork* (Unpublished thesis). University of Exeter, Exeter.

Wohlleben, P. (2016) *The hidden life of trees: What they feel, how they communicate – Discoveries from a secret world*. London: William Collins.

9 Opening spaces for the appearance of new subjectivities, action and hope

In this book I have argued for sustainable and democratic education which has the potential to enlarge '*the space of the possible*' (Davis *et al.* 2004: 4) rather than replicate the existing possible. Such sustainable and democratic education is *not* education which limits itself to sustainability and democracy as 'topics' of study, or school-based activities modelled on existing Western democratic approaches. Whilst these can have value, they are not *all* that education can and should be. Rather, the sustainable and democratic education I advocate requires a *revisioning* of education, particularly education influenced and constrained by the emphasis in dominant Western (Eurocentric) philosophy on static framings of the world and rational autonomous conceptions of subjectivity.

Such a re-imagining is towards educational processes which *can* include learning about and experiencing existing ideas, particularly when these are drawn from *many and various* framings of the world. However, there also needs to be opportunities to encourage the opening of spaces for playful engagement with abundant ideas and situations, for encountering others, for allowing emergence of the new, including new subjectivities, and for possibilities of unforeseen and unforeseeable futures. Education which encourages such novelty is democratic education when democracy is understood, as in Arendt's thinking, as freedom. This is freedom to speak and act with others, to be open to the stance they express, to respond in unexpected ways and to open spaces in and through which new ways of being and living in the world can appear. Such a revisioning is needed now more than ever before, in this era of the Anthropocene, where Western hubris threatens both human survival and that of myriad others on our shared planet.

Being open to the stance the other expresses is challenging, even overwhelming. It requires a critical exploration of one's existing framings of the self and the world – a process Spivak (1993) calls doing one's 'homework'. This requires perseverance and critical questioning. Using models such as HEADS UP (Andreotti 2012, 2015, 2016) can help, enabling a shift from a sense of 'drowning' to a sense of diving into 'creative potential' (*Gesturing toward Decolonial Collective*, online).

Intersubjective first-person and first-being encounters are important for the opening of spaces of appearance in and through which unique subjectivities can emerge. Such spaces can be encouraged, though never be guaranteed, through playfulness, listening, attentiveness, entanglement, entwining, tenderness towards the other and a willingness to enter into 'the space between I and we' (Topolski 2015: 176). Extending beyond the thinking of Arendt, whose focus is the human realm, first-being encounters engage with the possibilities of subjectivity in the wider natural world, what this can mean for sustainable and democratic education and why this is important in the era of the Anthropocene. Encouraging first-being, intersubjective encounters involves 'carefully listening to available voices' and requires 'actively de-centring the taken-for-granted human voice and the re-centring of more-than-human voices' (Jickling 2018: 35).

Encouraging emergence of the new can be problematic, raising questions such as what happens if what emerges is not seen as desirable and who gets to decide. Arendt's two-fold concept of forgiveness and mutual promising provides a way to approach the ethical issues arising from the unboundedness, irreversibility and unexpectedness of one's new beginnings when they are inserted into the world and taken up by others. This is an immanent approach to ethics which arises horizontally in and through speaking and acting together. Forgiving and making mutual promises, rather than resorting to vengeance, can be an unexpected act and thus an expression of freedom – a democratic move. Developing immanent ethical approaches can also be supported by learning with and from Indigenous conceptions of relationality, feminist ethics of care and Bennett's ideas of enchantment and the energy it can generate for responding ethically.

This is a hopeful book. This hope is not what Dryzek (2005) calls a 'Promethean' view in which we rely on human ingenuity to find ways for us to continue with 'business as usual'.[1] Instead it is a hope founded on approaches which challenge the myth of separateness whilst also not erasing the uniqueness of each participant in the world: a hope which values all humans and parts/ participants of the wider natural world, as well as the inter and intra connection between us. Having hope is not always easy. I am grateful to the Goorie-Koori poet Evelyn Araluen Corr, who shared with me that in her Indigenous language the concept of hope includes a sense of 'acting as though hope were possible'. This opens a way to act *for* and *towards* hope even in situations which appear hopeless: a situation many feel in the world today. It emphasises the necessarily *active* nature of hope – a focus on 'doing' rather than 'contemplation' – calling to mind Orr's (2007: 1) assertion that 'Hope is a verb with its sleeves rolled up'.

I have made arguments in which adults and children reposition themselves as actors in the world. Through speaking and acting with others, with all participants of the wider natural world, they have the potential to generate power to start and be/become something new, making possible the emergence of new ways of knowing, being and acting in the world. These possibilities can help to address the ecological crisis our shared planet faces and open unexpected futures. Education, in such a logic, has the potential to be a place where we:

> love our children [and students of all ages – *my addition*] enough not to expel them from our world and leave them to their own devices, nor to strike from their hands their chances of undertaking something new, something unforeseen by us.
>
> Arendt (2006a [1961]: 193)

Note

1 Dryzek (2005) develops a number of discourses or categories for the ways that different individuals or societies approach the issue of sustainability and possible responses to the challenges it poses.

References

Andreotti, V. (2012) Editor's preface: HEADS UP. *Critical Literacy: Theories and Practices*, 6(1): 1–3. Available at: www.oregoncampuscompact.org/uploads/1/3/0/4/13042698/andreotti_-_preface_-critical_literacy_org_-_headsup__1_.pdf [Accessed 23.1.2020].

Andreotti, V. (2015) Global citizenship education otherwise: Pedagogical and theoretical insights. In A. Abdi, L. Shultz and T. Pillay (eds.) *Decolonizing global citizenship education*. Rotterdam: Sense Publishers, pp. 221–230.

Andreotti, V. (2016) The educational challenge of imagining the world differently. *Canadian Journal of Development Studies/Revue Canadienne d'Etudes du Développement*, 37(1): 101–112.

Arendt, H. (2006a [1961]) Crisis in education. In *Between past and future: Eight exercises in political thought*. London: Penguin.

Davis, B., Phelps, R. and Wells, K. (2004) Complicity: An introduction and a welcome. *Complicity: An International Journal of Complexity and Education*, 1(1): 1–7.

Dryzek, J. (2005) *The politics of the earth: Environmental discourses,* 2nd Edition. Oxford: Oxford University Press.

Jickling, B. (2018) *On wilderness.* In R. Jickling, S. Blenkinsop, N. Timmerman and M. De Danann Sitka-Sage (eds.) *Wild pedagogies: Touchstones for re-negotiating education and the environment in the Anthropocene* (Palgrave studies in educational futures). New York: Palgrave Macmillan, pp. 23–50.

Orr, D. (2007) Optimism and hope in a hotter time. *Conservation Biology*, 21(6): 1392–1395. doi: 10.1111/j.1523-1739.2007.00836.x

Spivak, G. (1993) *The post-colonial critic: Interviews, strategies, dialogues.* New York: Routledge.

Topolski, A. (2015) *Arendt, Levinas and a politics of relationality* (Reframing the boundaries: Thinking the political). London: Rowman & Littlefield International.

Appendix
Drawing on case studies as starting points for 'enlarging the space of the possible' in educational settings

Case studies can be used in a number of ways in educational settings, for example to provide opportunities to apply learning, to broaden thinking and to encourage further research. Whilst they can be used in closed ways in which the teacher already has a range of pre-set questions and often an idea of the 'correct solution' they can also be used in more open ways as starting points for independent thinking and to encourage the complex interactions and connections necessary for emergence. The case studies included in this book are intended as such starting points and can be used in a variety of ways, including ones not outlined here! In highly regulated educational environments, it can be hard to believe that 'something will emerge', but teachers and students can develop confidence and take what might be tentative steps at first to try this out and to share the experiences opened up.

Ideas for using case studies as starting points include:

- Sharing the case studies in this book with students as starting points for reflection, discussion and development of further learning opportunities. Students can be involved in developing exploratory questions to support this process.
- Educators reading the case studies and then using these to develop ideas for learning inside and outside the classroom, for example opportunities for intersubjective first-person and first-being encounters. Ideas discussed in this book such as touchstones for wild pedagogies, opportunities for embodied experiences and the 'lingering, mingling, exploring, noticing, connecting, sharing and greeting a multitude of things flourishing in relation' (Witt 2019) can be drawn in.
- Using the case studies as initial 'topic' areas for students to carry out their own research (e.g. international responses to climate change) without the teacher being 'overly' prescriptive about the direction taken. For example, in relation to international responses to climate change some students might choose to research United Nations activities, as in this book, whilst others might wish to research food production, individual government actions, political and citizen resistance, etc.

- Discussing ambiguity with colleagues and students, including how comfortable one is with ambiguity.
- Considering a variety of ways that ideas can be explored and responded to. This could include aesthetic engagements such as music, poetry, song, drama, dance and art which take learning beyond cognitive activities and towards more embodied responses. Educators do not need to take responsibility for devising aesthetic methods. Students can choose and design their own method. This is appropriate since it is their particular emergent, aesthetic, embodied response which is being encouraged.

Good luck and enjoy!

Reference

Witt, S. (2019) *Becoming lost in relational, democratic geographical fieldwork* (Unpublished thesis). University of Exeter, Exeter.

Glossary of terms

Agonistic pluralism   This is a term used by Mouffe (2000, 2005) as a desirable alternative to antagonistic pluralism. Agonistic pluralism allows for a recognition that there will always be those whose views are different from one's own, whose preferred courses of action are equally valid but not chosen and that such a situation is an inevitable part of the human condition. It recognises that those holding different views from one's own can be adversaries to be defeated but respected rather than enemies to be destroyed, silenced or converted, or treated as though they did not exist in the first place.

Anthropocene   A term used widely since its coining by Paul Crutzen and Eugene Stoermer in 2000 to denote the present time interval, in which many geologically significant conditions and processes are profoundly altered by human activities. These include changes in: erosion and sediment transport associated with a variety of anthropogenic processes, including colonisation, agriculture, urbanisation and global warming, the chemical composition of the atmosphere, oceans and soils, with significant anthropogenic perturbations of the cycles of elements such as carbon, nitrogen, phosphorus and various metals (International Commission on Stratigraphy 2016). In 2016 the International Commission on Stratigraphy working party voted in favour of accepting the term, subject to identification of a specific signal which can mark the change.

Borderlands/border thinking   Borderlands/border thinking is a term developed and explored in decolonial theory (see Anzaldúa 1999, Mignolo and Tlostanova 2006). It is based on the ideas that 'the theoretical and epistemic must have a lived dimension to them' and 'that theories already exist which sit at the very borders (if not outside of) the colonial matrix of power' (globalsocialtheory.org: online).

Complexity-compatible thinking   Complexity-compatible thinking is a term used to describe theories and ideas which, whilst not identified as complexity theory, are compatible with it and sometimes share some of its historical sources (Osberg 2015).

Democratic education   In this book the term democratic education refers to education which opens space for unique subjects to respond in unexpected ways to situations one faces and to bring new possibilities and ways to be/become into the world.

Dissensus   Rancière (2010: 69) uses the term *dissensus* not to denote a 'conflict of interests, opinion, or value'. Instead *dissensus* is 'a division inserted in "common sense"; a dispute over what is given and about the frame within which we see something as given'.

Ecocide   Ecocide is such extensive damage to, destruction of or loss of ecosystem(s) that the very survival of that ecosystem and all who form part of it is threatened. Whilst not yet recognised by the United Nations as an international crime, organisations such as STOPECOCIDE are 'campaigning for it to be recognised as an atrocity crime at the International Criminal Court – alongside Genocide, War Crimes and Crimes Against Humanity' (STOPECOCIDE: online).

(European) Enlightenment   The (European) Enlightenment was a seventeenth- and eighteenth-century movement which emphasised universal Reason. It was a reaction against the notion that 'the people' were incapable of understanding the world for themselves and indeed did not need to understand but just 'obey their betters' (God, the church, the Sovereign, the aristocracy). 'The people' were thus not free. The Enlightenment argued that freedom from this position could be attained through Reason and that each individual had the capacity to use their reason autonomously. There were limits to who was deemed capable of such Reason, however, as certain categories of individual, for example women and children, were excluded by some thinkers. The (European) Enlightenment had a wideranging influence on political and cultural life, from literature to political revolution.

(European) Renaissance   The Renaissance, literally 'rebirth' is a term used to refer to the 'the period in European civilization immediately following the Middle Ages and conventionally held to have been characterised by a surge of interest in Classical [Ancient Greek] scholarship and values' after what the European scholars and thinkers of that time identified as a 'long period of cultural decline and stagnation'. Broadly speaking, the Middle Ages refers to the period beginning in the fifth century and the Renaissance is considered to have commenced in the thirteenth, fourteenth or fifteenth century depending on area of Europe and other factors (Encyclopaedia Britannica 2020: online).

First–person intersubjective encounter   A first-person intersubjective *encounter* is an encounter with other humans in which, through speech and action with others where each is open to the stance the other expresses, *who* one is as a unique subject can emerge 'intersubjectively (i.e. arising in and through the encounters rather than preceding it).

First-being intersubjective encounter   A first-being intersubjective *encounter* or event is an encounter in which a subject can express their uniqueness to the other-than-human who is (potentially) open to its expression and can express their own uniqueness. In such encounters *who* a person is as a unique being has the potential to emerge intersubjectively (between subjects).

Global North and Global South   The terms Global North and Global South were introduced in the 1990s, and they became increasingly popular from 2000 onwards. They are based on the United Nations Development Programme's (UNDP) Human Development Index (HDI) to differentiate between different parts of the world. The term is generally understood as follows: the Global North consists of those 64 countries which have a high HDI (most of which are located north of the 30th parallel north), whilst the remaining 133 countries belong to the Global South. For a critique of these terms see Hylland Eriksen (2015).

Hyperseparation   Plumwood (2001) uses this term to refer to the two-fold separation in Western (Eurocentric) framings of the human from other parts/participants in the wider natural world. First, the human is separate because it is part of a separate species from these others. Second, the human is separate because in Western constructions the human is also placed as superior to and dominant over other species.

Initium   In Arendt's thinking, an *initium* is a beginner who can open up new ways of knowing, being and acting in the world.

Intersubjectivity   There are a variety of understandings of intersubjectivity in philosophical thinking. In this book I develop an immanent understanding of intersubjectivity in which *who* one is as a subject *enacts* itself in the world in and through intersubjective first-person/first-being encounters with others unlike itself where each is open to the stance the other expresses.

[See also earlier entries on first-person and first-being intersubjective encounters.]

Modernism   A movement which began to emerge in the Western world in the late Medieval and early Renaissance period in Europe and continued to develop in the centuries which followed. It came to the fore in the mid-nineteenth century and continues to be influential in the present time. It deliberately rejected ideas of the past and emphasised rationality, innovation and scientific development in an era of increasing industrialisation and urbanisation.

Natality   Arendt's notion of natality has a sense of both birth and 'second birth' and is a way that humans have the potential repeatedly to bring new ways of being into the world through words and deeds.

Plurality (e.g. under conditions of plurality)   Drawing on Arendt's thinking the phrase 'under conditions of plurality' refers to encounters when there is opportunity to speak and act with others where each is open to the stance the other expresses (see discussion in Chapter 4).

Radical   In this book I use the word 'radical' to identify something which is 'uniquely new, something which has not been in the world before, and cannot be predicted from the ground from which it emerged' (Osberg and Biesta 2008: 313).

Skholé   A break from the issues of the day, 'free time' to encourage the possibility of the new. Literally translated from the Ancient Greek *skholé* means free time, a break, a respite, leisure. However, for the Ancient Greeks, this was not leisure in the sense of a luxury or break from a primary activity. Rather it denotes a time which has a higher value than what it is interrupting, a time to debate and to reflect.

   The word *skholé* began to be used to refer to the site where such activities took place, rather than the processes themselves, and thus became the etymological root of the noun 'school'.

Spaces of appearance   Spaces where, through acting and speaking in the presence of others who are themselves unique beings, 'who' one is as an *initium*, a beginner can emerge, bringing new ways of knowing, being and acting into the world.

Subjectivity   In the Western philosophical tradition 'subjectivity' refers to a sense of self: a sense of who one is, how one acts in the world and a capacity to have higher order thoughts and reflect on selfhood. Recent thinking has explored possibilities that other ways to theorise and also experience subjectivity are possible, thus broadening possibilities of who can have subjectivity (see discussion in Chapter 8).

Sustain   'To cause something to continue for an extended period of time' (Oxford Living Dictionaries: English: online). This can include a wide range of parts/participants in the world, for example plants, animals, rock, water and sand as part of a landscape and the ecosystems integral to it.

Sustainability   'The capacity to endure. In ecology the word describes how biological systems remain diverse and productive over time. For humans it is the potential for long-term maintenance of well-being which in turn depends on the natural world and natural resources' (Habitat.org.tr 2016: online).

Sustainable development   'Humanity has the ability to make development sustainable – to ensure that it meets the needs of the present without compromising the ability of future generations to meet their own needs' (WCED 1987: 1). The emphasis on development is contested by some, although others argue that the development needed for poverty eradication also needs to be taken into consideration in discussions over sustainability (for further discussion see Dresner 2002).

Western (Eurocentric) philosophy/Western philosophical tradition   Western philosophy refers to philosophical thinking in the Western world (beginning with Ancient Greece and Rome, extending through central and Western Europe and, since Columbus, the Americas). Since it builds on Ancient Greek ideas, it is sometimes referred to as Hellenic philosophy.

Wider natural world   I use this term when I talk about the other-than-human to indicate that humans are also a part of the natural world. Some choose to use the phrase 'more' than human to indicate both a 'beyond the human' and also challenge/reverse the privileging of humans in Western cultures in which the human is placed above other participants of the natural world. However, I prefer the term wider natural world as it does not introduce or reinforce hierarchies of any kind.

References

Anzaldúa, G.E. (1999) *Borderlands/La frontera: The new Mestiza*. San Francisco: Aunt Lute Books.

Dresner, S. (2002) *The principles of sustainability*. London: Earthscan.

Encyclopaedia Britannica (2020) *Renaissance*. Available at: https://www.britannica.com/event/Renaissance [Accessed 24.9.2020].

Globalsocialtheory.org (online) *Border thinking*. Available at: https://globalsocialtheory.org/concepts/border-thinking/ [Accessed 11.9.2019].

Habitat.org.tr (2016) *Sustainability*. Available at: http://environment-ecology.com/what-is-sustainability/247-sustainability.html [Accessed 12.11.2016].

Hylland Eriksen, T. (2015) *What's wrong with the Global North and the Global South: Voices from around the world. Cologne*. Global South Studies Centre. Available at: http://gssc.uni-koeln.de/node/454 [Accessed 3.9.2016].

International Commission on Stratigraphy (2016) *Subcommission on quaternary stratigraphy: Working group on the 'Anthropocene'*. Available at: http://quaternary.stratigraphy.org/workinggroups/anthropocene/ [Accessed 10.11.2016].

Mignolo, W.D. and Tlostanova, M.V. (2006) Theorising from the borders: Shifting to geo- and body-politics of knowledge. *European Journal of Social Theory*, 9(2): 205–221.

Mouffe, C. (2000) *Deliberative democracy or agonistic pluralism?* (Political science series). Vienna: Institute for Advanced Studies. Available at: www.ihs.ac.at/publications/pol/pw_72.pdf [Accessed 2.12.2016].

Mouffe, C. (2005) *The democratic paradox*. New York: Verso.

Osberg, D. (2015) Learning, complexity and emergent (irreversible) change. In E. Hargreaves (ed.) *Sage handbook of learning*. London: Sage, pp. 23–40.

Osberg, D. and Biesta, G. (2008) The emergent curriculum: Navigating a complex course between unguided learning and planned enculturation. *Journal of Curriculum Studies*, 40(3): 313–328. doi: 10.1080/00220270701610746

Oxford Living Dictionaries: English (online) *Sustain*. Oxford Living Dictionaries: English. Available at: https://en.oxforddictionaries.com/definition/sustain [Accessed 11.1.2017].

Plumwood, V. (2001) The concept of a cultural landscape: Nature, culture and agency in the land. *Ethics and the Environment*, 11(2): 115–150.

Rancière, J. (2010) Who is the subject of the rights of man? In J. Rancière (ed.) *Dissensus: On politics and aesthetics*. London: Continuum, pp. 70–83.

STOPECOCIDE (online) *Making ecocide a crime*. Available at: www.stopecocide.earth/making-ecocide-a-crime [Accessed 30.1.2020].

World Commission on Environment and Development (WCED) (1987) *Our common futures* (Brundtland Report). Available at: www.un-documents.net/our-common-future.pdf [Accessed 10.11.2016].

Index